PHYTOREMEDIATION

78

Advances in Biochemical Engineering / Biotechnology

Managing Editor: T. Scheper

Springer
Berlin
Heidelberg
New York
Hong Kong
London
Milan
Paris
Tokyo

Phytoremediation

Volume Editor: David T. Tsao

With contributions by
M. K. Banks, A. Ferro, F. W. Flechas, M. Gefell, S. Jackson,
R. Karthikeyan, R. Kim, R. Kjelgren, P. A. Kulakow,
M. Latady, D. S. Lipson, B. Liu, T. McIntyre, S. A. Rock,
P. Schwab, J. S. Smith, D. T. Tsao, N. Zollinger

Springer

Advances in Biochemical Engineering/Biotechnology reviews actual trends in modern biotechnology. Its aim is to cover all aspects of this interdisciplinary technology where knowledge, methods and expertise are required for chemistry, biochemistry, microbiology, genetics, chemical engineering and computer science. Special volumes are dedicated to selected topics which focus on new biotechnological products and new processes for their synthesis and purification. They give the state-of-the-art of a topic in a comprehensive way thus being a valuable source for the next 3–5 years. It also discusses new discoveries and applications.

In general, special volumes are edited by well known guest editors. The managing editor and publisher will however always be pleased to receive suggestions and supplementary information. Manuscripts are accepted in English.

In references Advances in Biochemical Engineering/Biotechnology is abbreviated as *Adv Biochem Engin/Biotechnol* as a journal.
Visit the ABE home page at http://link.springer.de/series/abe/
http://link.Springer-ny.com/series/abe/

ISSN 0724-6145
ISBN 3-540-43385-6
Springer-Verlag Berlin Heidelberg New York

Library of Congress Catalog Card Number 72-152360

Springer-Verlag Berlin Heidelberg New York
a member of BertelsmannSpringer Science+Business Media GmbH

http://www.springer.de

Typesetting: Fotosatz-Service Köhler GmbH, Würzburg
Cover: E. Kirchner, Heidelberg

Printed on acid-free paper 02/3020mh – 5 4 3 2 1 0

Managing Editor

Professor Dr. T. Scheper
Institute of Technical Chemistry
University of Hannover
Callinstraße 3
30167 Hannover, Germany
E-mail: scheper@mbox.iftc.uni-hannover.de

Volume Editor

Prof. David T. Tsao
Hydrocarbon and Environmental Management
Group Global Environmental Management
Business Unit
150 W. Warrenville Rd.
Naperville, IL 60563, USA
E-mail: tsaodt@bp.com

Editorial Board

Advances in Biochemical Engineering Biotechnology now also Available Electronically

For all customers with a standing order for Advances in Biochemical Engineering/Biotechnology we offer the electronic form via LINK free of charge. Please contact your librarian who can receive a password for free access to the full articles. By registration at:

http://www.springer.de/series/abe/reg_form.htm

If you do not have a standard order you can nevertheless browse through the table of contents of the volumes and the abstracts of each article at:

http://link.springer.de/series/abe/
http://link.springer_ny.com/series/abe/

There you will find also information about the

- Editorial Board
- Aims and Scope
- Instructions for Author

Attention all Users of the Springer Handbook of Enzymes

Information on this handbook can be found on the internet at
http://www.springer.de/enzymes/

A complete list of all enzymes entries either as an alphabetical Name Index or as the EC-Number Index is available at the above mentioned URL. You can download and print them free of charge.

A complete list of all synonyms (more than 25,000 entries) used for the enyzmes is available in print form, ISBN 3-540-41830-X.

Save 15%

We recommend a standing order for the series to ensure you automatically receive all volumes and all supplements and save 15% on the list price.

Preface

The environmental clean up industry has been estimated as having an annual turnover of $50 billion globally. With new regulations being written on additional chemicals that are just, now, becoming understood from a toxicological and environmental risk standpoint, this industry could expand even further. This is particularly true as more nations become industrialized. Typical contaminants that are of concern include agricultural byproducts, municipal wastes, industrial solvents, petroleum hydrocarbons, heavy metals, pesticides, radioactive wastes, munitions, and other man-made products.

In order to treat and remediate these contaminants, practioners have several "tools" in the remediation "toolbox" including physical, chemical, and biological methods. One relatively new biological method that has been applied to address various environmental concerns is phytotechnologies. The method is defined as the use of vegetation to contain, sequester, remove, or degrade inorganic and organic contaminants in soils, sediments, surface waters, and groundwater. Although its roots were developed from other disciplines such as agronomy, agricultural engineering, chemical engineering, forestry, horticulture, hydrogeology, and microbiology, this set of technologies has grown substantially on its own in understanding of and application in the environmental clean-up industry around the world.

This broad-ranging set of technologies utilizes the complex processes occurring within the soil-plant-atmosphere continuum in order to clean up and restore environmentally impacted sites. Like all remediation technologies, the use of vegetation is appropriate under specific situations, but it can be utilized to address both organic and inorganic constituents as well as applied to remediating different impacted media. In some cases, it can address many of these situations simultaneously. This and its broad applicability are, perhaps, its biggest advantage in addition to some other ancillary benefits including aesthetics, waste minimization, low energy requirements, erosion control, greenhouse gas emissions reduction, and cost-effectiveness.

This edition of Advances in Biochemical Engineering and Biotechnology provides a general overview of phytotechnologies. Authors from private industry, academia, the consulting community and regulatory agencies have all contributed to this volume. Starting with the basic processes and mechanisms occurring in the soil-plant-atmosphere continuum, the typical applications of vegetation for cleaning up and remediating contaminated sites are described. Since the growth of vegetation is paramount to this technology, the factors within the subsurface environment affecting growth are discussed in detail as well. These include the processes and factors in the soil environment that affect the plant and the subsurface microbial community. After these two "introductory" chapters, the most prevalent applications of phytotechnologies are reviewed. These include vegetated systems for treating organic and inorganic contaminants in solid media, hydraulic systems for controlling and treating contaminant groundwater plumes, and vegetative covers for surface water protection and landfill leachate management. Finally, this monograph concludes with a discussion of the regulatory implications of applying this technology towards the clean up and remediation of contaminated sites worldwide.

Naperville, August 2002 David T. Tsao

Contents

Overview of Phytotechnologies

David T. Tsao

Group Environmental Management (aBP – affiliated company), 150 W. Warrenville Rd., Mail Code *H*-7, Naperville, IL 60563, USA. *E-mail: tsaodt@bp.com*

Phytotechnologies are defined as:

> "The use of vegetation to contain, sequester, remove, or degrade inorganic and organic contaminants in soils, sediments, surface waters, and groundwater."

This chapter is designed to provide the reader with a broad, introductory overview of phytotechnologies as well as environmental remediation in general. To familiarize the reader with the common acronyms, abbreviations, and parameters used in the environmental arena, an extensive list is provided for quick reference. Furthermore, due to the multi-disciplinary nature of phytotechnologies, a brief background on the pertinent information in plant physiology, soil microbiology, hydrogeology, and environmental engineering is provided as well. This covers the basic physiological processes occurring in the rhizosphere as well as in the plant system. Furthermore, these basic processes are related to the phytotechnology mechanisms that form the basis for the various applications used in the field today. Finally, basic design equations, site-specific considerations, and practical field implementation techniques are provided as well for the different applications.

Keywords. Rhizosphere, Plant physiology, Contamination, Phytoremediation, Treatment, Containment

Advances in Biochemical Engineering/
Biotechnology, Vol. 78
Managing Editor: T. Scheper

Abbreviations

2,4-D	2,4-Dichlorophenoxyacetic Acid
A	Area of the Phytotechnology System (ft^2 or acres)
ATPase	Adenosine Triphosphatase
bgs	Below Ground Surface
BOD_5	Biochemical Oxygen Demand (5 day)
BTEX	Benzene, Toluene, Ethylbenzene, (o-, m-, p-) Xylenes
cfu	Colony Forming Units
COD	Chemical Oxygen Demand
C_p	Specific Heat of Air (J/kg °C)
C_w^{sat}	Solubility (mol/L)
D	Vapor Pressure Deficit (Pa)
DCE	Dichloroethylene
D_p	Depth to Plume (ft)
DRO	Diesel Range Organics
D_r	Depth of Root System (ft)
D_t	Thickness of Plume (ft)
D_u	Depth of Influence of Root Systems into a Saturated Zone (ft)
EDTA	Ethylene Diamine Tetraacetic Acid
ET	Evapotranspiration (gallons per day)
f	fraction of Rain Available to Infiltrate into Soil (%)

FAC	Facultative Species (Wetland Indicator Status)
FACU	Facultative Upland Species (Wetland Indicator Status)
FACW	Facultative Wetland Species (Wetland Indicator Status)
f_R	Retardation Factor (dimensionless; less than one)
G	Soil Heat Flux Density (W/m^2)
g_a	Aerodynamic Conductance (m/s)
g_c	Canopy Conductance (m/s)
GRO	Gasoline Range Organics
HELP	Hydrologic Evaluation of Landfill Performance
HMX	Octahydro-1,3,5,7-tetranitro-1,3,5,7-tetraazocine
i	Fraction of Rain Intercepted by Vegetation (%)
I	Rate of Infiltration (gallons per day)
K_d	Phase Distribution or Partitioning Coefficients (dimensionless)
K_H	Henry's Law Gas Constant (dimensionless)
LAI	Leaf Area Index (dimensionless)
$\log K_{ow}$	Octanol-Water Partition Coefficient (dimensionless)
MPN	Most Probable Number
MTBE	Methyl Tertiary Butyl Ether
NPDES	National Pollution Discharge Elimination System
OBL	Obligate Wetland Species (Wetland Indicator Status)
O&G	Oil and Grease
P_0	Vapor Pressure (atm)
PAH	Polycyclic Aromatic Hydrocarbons
PCB	Polychlorinated Biphenyls
PCE	Perchloroethylene (Tetrachloroethylene)
PCP	Pentachlorophenol
PCR-DGGE	Polymerase Chain Reaction – Denaturing Gradient Gel Electrophoresis
PET	Potential Evapotranspiration (kg/m^2s)
PLFA	Phospholipid Fatty Acid
POTW	Publicly Operated Treatment Works
ppm	Parts Per Million
Q	Groundwater/Surface Water Flow Rate (gallons per day)
r	Fraction of Rain Forming Run-Off (%)
R	Volumetric Rate of Rainfall per Unit Area (inches/day or mm/day)
R′	Volumetric Rate of Rainfall (gallons per day)
RCF	Root Concentration Factor (µg/kg root per µg/L)
RCRA	Resource Conservation and Recovery Act
rDNA	Recombinant DNA
RDX	Hexahydro-1,3,5-trinitro-1,3,5-triazine
R_n	Net Solar Radiative Flux Density (W/m^2)
SF	Surface Flow Wetland System
SSF	Subsurface Flow Wetland System
SVOC	Semi-Volatile Organic Compounds
TCA	Trichloroethane
TCAA	Trichloroacetic Acid
TCE	Trichloroethylene

TCH	Total Culturable Heterotrophs
TDS	Total Dissolved Solids
TKN	Total Kjeldahl Nitrogen
TNT	Trinitrotoluene
TOC	Total Organic Carbon
TPH	Total Petroleum Hydrocarbons
TSCF	Transpiration Stream Concentration Factor (dimensionless)
TSS	Total Suspended Solids
UPL	Obligate Upland Species (Wetland Indicator Status)
VAM	Vesicular-Arbuscular Mycorrhizal
VC	Vinyl Chloride
v_C	Seepage Velocity for Groundwater Contaminants (ft/day)
v_D	Darcy Velocity for Groundwater (ft/day)
VF	Vertical Flow Wetland System
VOC	Volatile Organic Compounds
V_T	Vegetative Water Uptake (gallons per day)
W	Width of a Groundwater Plume (ft)
WWTP	Wastewater Treatment Plant
z	Water Depth of Wetland Treatment System (ft)
γ	Psychrometric Constant (Pa/°C)
Δ	Rate of Change in the Saturation Vapor Pressure (Pa/°C)
$\Delta G°$	Gibbs Free Energy Changes (kcal/mol)
λ	Latent Heat of Vaporization (J/kg)
ϱ	Density of Air (kg/m^3)

1 Introduction

Phytotechnologies are defined as:

> "The use of vegetation to contain, sequester, remove, or degrade inorganic and organic contaminants in soils, sediments, surface waters, and groundwater."

The first part of this sentence, "The use of vegetation to contain, sequester, remove, or degrade" implies that this technology involves several complex processes that form the mechanistic basis behind phytotechnologies. The functioning processes by soil organisms includes protein and enzyme production, bio-catalysis, and symbiosis with plant roots. Similarly, the processes occurring within the plant system that have been exploited in phytotechnologies include inorganic nutrition, water uptake, evapotranspiration, photosynthetic production of phytochemicals, root exudation, root turnover, and bio-metabolism. An integral component of these processes is the interaction that occurs between plants and microbes at the root-soil interface known as the rhizosphere. This region of soil as depicted in Fig. 1 is approximately 1–3 mm surrounding individual roots and is highly bioactive [1, 2]. Several of the mechanisms that form the basis for phytotechnologies focus primarily on the processes occurring within this zone. Therefore, a basic understanding of the operating processes occurring in the

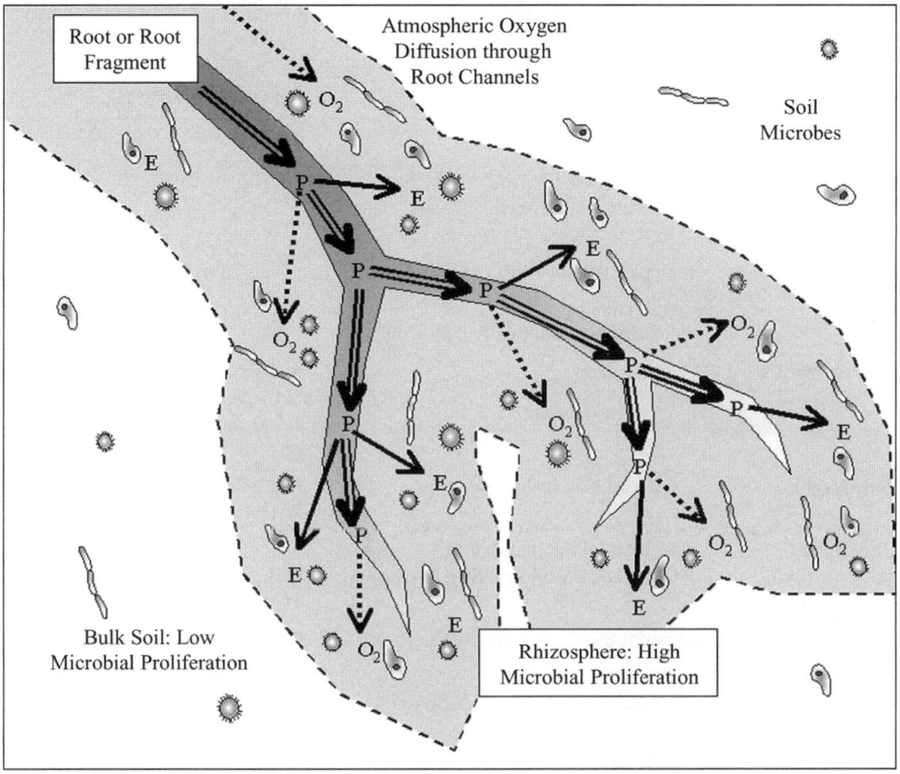

Fig. 1. Plant root rhizosphere (P ⇒ P = phytochemicals produced through photosynthesis and translocated through the phloem, P → E = phytochemicals released as exudates, P --→ O_2 = phytochemicals consumed during respiration to produce released oxygen)

rhizosphere is necessary in order to define and utilize the mechanisms of phytotechnologies.

The second part of the definition, "inorganic and organic contaminants" implies that phytotechnologies can be applied to a wide range of chemicals that are concerns in the environment. The specific types of inorganic contaminants that have been addressed using phytotechnologies include heavy metals, metalloids, radionuclides, salts, and agrochemicals. For organic contaminants, hydrocarbons, chlorinated compounds, and nitroaromatics have been treated as well. In addition, facilities where mixed wastes usually reside, such as landfills and wastewater treatment facilities, have utilized these technologies as well. Several example constituents and common sources are provided in Table 1. The specific mechanism that is incorporated into the final design of a phytotechnology system is dependent on the type of constituent that needs to be addressed and the clean up objectives for the site. Certain inorganics can be sequestered in a containment strategy while others can be extracted if the goal is remediation. Similarly, dissolved organic compounds can be contained by controlling the subsurface hydraulics through vegetation while remediation can be achieved by volatilizing or de-

Table 1. Environmental contaminants and common sources addressed using phytotechnologies

Contaminant class	Common sources	Specific constituents
Heavy Metals	Mining Sites, Smelters	Ag, Au, Cd, Cr, Cu, Mn, Ni, Pb, Zn
Metalloids	Power Plants, Wood Treating Facilities, Agriculture	As, Hg, Se
Radionuclides	Department of Energy Facilities	^{137}Ce, ^{239}Pu, ^{90}Sr, $^{234/238}$U, tritium
Salts	Agriculture, Oil and Gas Production, Metals Processing	Na, Ca, Mg Chlorides
Agrochemicals	Agriculture	Atrazine, Metolachlor, (Pesticides, Fertilizers) Propanil, 2,4-D, NO_3, NH_4, TKN, Phosphorus, Organophosphates
Hydrocarbons	Oil and Gas Industry	BTEX, MTBE, GRO, DRO, PAH, O&G, VOC, SVOC, TPH
Chlorinated Compounds	Industrial Facilities, Dry-Cleaners, Wood Treating Facilities	PCE, TCE, DCE, VC, TCA, TCAA, PCP, PCB
Nitroaromatics	Department of Defense Facilities	Nitrobenzene, TNT, RDX, HMX (unexploded ordnance)
Leachate	Landfills (Municipal, Industrial)	Municipal, Hazardous, RCRA Wastes
Wastewater	WWTP, POTW	BOD_5, COD, TDS, TSS, TOC, phenols, *E. coli*, fecal coliform, pathogens

grading the contaminant. Therefore, a clear understanding of the phytotechnology mechanisms allows for specific applications to be engineered in order to meet the desired clean up goals.

The third portion of the defining sentence, "in soils, sediments, surface waters, and groundwater" indicate that this technology is also applicable to a whole host of impacted media. As with the type of contaminant, the type of the impacted media also dictates the application that can be utilized. The major vegetated structures that are developed include tree stands, grass and herbaceous plots, riparian transitions, and wetland areas. Furthermore, since vegetation is classified based on the indicator status including obligate wetland (OBL), facultative wetland (FACW), facultative (FAC), facultative upland (FACU), and obligate upland (UPL) species, then the class of vegetation that can be used in the system directly affects the design of the application. In general, UPL species are used to treat impacted soils while FAC and FACU species are applicable for groundwater treatment. Similarly, FAC and FACW species are useful for treating sediments or creating riparian zones while OBL species are characteristic of saturated (wetland) systems. Therefore, a basic understanding of the impacted media and classes of plants that corresponding to that media, can be used to design appropriate phytotechnology applications.

In addition to understanding the basic physiological processes, mechanisms, and engineered applications, several other factors that affect the overall use of phytotechnologies need to be considered as well. These include the economics, capital and operation and maintenance requirements, monitoring needs, public perception, and regulatory environment. Furthermore, the proper phytotechnology system must be designed, developed, and implemented using detailed knowledge of the site layout, geology, hydrology, climate conditions, botany, horticulture, forestry, and engineering. Therefore, phytotechnologies is a highly cross-disciplined field requiring a broad range of expertise.

2 Basic Physiological Processes

The subsurface environment in the vicinity of vegetation is highly bioactive. Plants and soil microbes have evolved highly complex symbiotic and synergistic relationships that provide the plant with protection, nutrition, and enhanced water uptake capacities while the soil microbes are provided with an enhanced nutritional environment from which to thrive. This region of soil is known as the rhizosphere and contains multiple biological processes that are pertinent to phytotechnologies. Furthermore, additional processes occurring strictly within the plant, but that affect the subsurface environment, are discussed as well.

2.1 Microbial Processes

Microbes exist ubiquitously in the subsurface environment. This includes bacteria, actinomycetes, molds, fungi, algae, viruses, and protozoa. These organisms have evolved into a particular community structure utilizing the carbon naturally available in the subsurface. Typical concentrations in non-vegetated soils (or bulk soils) at the surface (A Horizon) are in the range of 10^5 to 10^8 cells per gram of dry soil. As one progresses downward into the subsurface stratigraphy (B and C Horizons), the populations drop off substantially to about 10^3 to 10^6 cells/gram. Groundwater typically contains up to 10^5 cells per ml [3]. In comparison, rhizosphere soils have been reported to contain populations that are 1 to 2 orders of magnitude higher than bulk surface soils [4]. However, specific populations of certain culturable microorganisms have demonstrated dominance and then decreased dramatically as the contaminant becomes weathered [5]. Basically, after the contaminant release, the organisms most suitable to metabolize the parent compound(s) predominate but eventually give way to other strains as by-product are generated, creating food-sources for other microbes. Similarly, measures of diversity have also shown predominance by a few organisms giving way to a broader array of microbes in time. Another occurrence that has been demonstrated is the switch in predominance from aerobic to anaerobic microbes as electron acceptors are consumed [6].

A caveat to these findings is the limitation of estimating populations due to the relative few culturable microorganisms that exist in the subsurface. Typical most probable number (MPN) and total culturable heterotroph (TCH) techniques yield results (reported as colony forming units, cfu) that represent less than 10% of the total organisms in the subsurface [7]. These are typically Methanotrophs, Actinomycetes, Pseudomonads, *Escherichia coli*, etc. More advanced techniques such as phospholipid fatty acid (PLFA) analysis, polymerase chain reaction – denaturing gradient gel electrophoresis (PCR-DGGE), and recombinant DNA (rDNA) sequencing have yielded better characterizations both quantitatively and qualitatively as well as increased accuracy in determining viable microbial biomass [8].

2.1.1 *Bio-Catalysis*

The different types of organisms present in the subsurface can be classified into aerobic, anaerobic, and facultative organisms [6]. Aerobic organisms only utilize oxygen as the terminal electron acceptor while anaerobes cannot use oxygen but utilize other electron acceptors instead. Listed in order of utilization, anaerobes cause the preferential donation of electrons from the contaminant to nitrate, manganese oxide, iron (III), and then sulfate. Once these are consumed, the contaminant can then act as the electron acceptor and receive electrons from carbon dioxide to produce methane during the process of methanogenesis. These are determined by the redox states of the electron donor – electron acceptor pair (see Gibbs free energy changes, $\Delta G°$, for each acceptor in Eq. (1)). In the absence of sufficient oxygen, aerobic organisms are incapable of metabolization; therefore, their function in the breakdown of contaminants becomes oxygen limited. However, some organisms are facultative in that they can utilize oxygen when in sufficient concentrations or they can switch to other acceptors under low or non-existent O_2 conditions.

$$O_2 \rightarrow NO_3^- \rightarrow MnO_2 \rightarrow Fe(III) \rightarrow SO_4^{2-} \rightarrow CO_2 \qquad (1)$$

$$\Delta G° \; O_2 = -120 \text{ kcal/mol}$$
$$\Delta G° \; NO_3^- = -114 \text{ kcal/mol}$$
$$\Delta G° \; MnO_2 = -81 \text{ kcal/mol}$$
$$\Delta G° \; Fe(III) = -28 \text{ kcal/mol}$$
$$\Delta G° \; SO_4^{2-} = -25 \text{ kcal/mol}$$
$$\Delta G° \; CO_2 = -21 \text{ kcal/mol}$$

In order for the soil organisms to thrive and become dominant, they must be able to tolerate and utilize the contaminant or a by-product in some adventitious manner. In general, these organisms can metabolize organic compounds (denoted as CH_2O in Eq. (2)) into a suitable carbon source by producing proteins and enzymes that can bio-catalyze the breakdown into basic components.

$$CH_2O + O_2 \rightarrow CO_2 + H_2O \qquad (2)$$

The rates of reaction are typically highest for aerobic biodegradation and decrease through the electron acceptor chain, see Eq. (1). For benzene, toluene, ethylbenzene, and (o-, m-, p-)xylenes (BTEX), the rates are generally in the range of 0.1 to 1.3 percent per day when modeled using first-order kinetics [9, 10]. In comparison, the anaerobic biodegradation of BTEX is in the range of 0.004 to 0.05 percent per day. Therefore, reducing oxygen limitations is advantageous from a clean up standpoint. One of the functions of plants in the symbiotic relationship with soil organisms is to maintain an oxygen-rich environment for optimum proliferation.

In addition to soil organisms that can directly metabolize organic contaminants, certain soil organisms can also cause the fortuitous co-metabolic biodegradation. For example, methanotrophic bacteria produce methane monooxygenase in the presence of oxygen and methane [11]. While consuming the methane, the enzyme is capable of co-metabolically degrading chlorinated solvents such as trichloroethylene (TCE). Similarly, several mono- and di-oxygenase enzymes derived from organisms that feed on aromatics such as phenols have been reported to be able to co-metabolically degrade TCE as well [6]. Even the relatively recalcitrant contaminant, methyl tertiary butyl ether (MTBE), has been shown to be co-metabolically biodegraded by a camphor-oxidizing bacteria [12]. Therefore, these phenolic compounds, which can be produced by plants, are capable of inducing these microbial enzymes to co-metabolically degrade various contaminants.

2.1.2
Symbiosis with Plant Roots

The soil organisms that breakdown organic contaminants provide protection for the plants by restricting contact with the potentially toxic compound. This protective "sheath" exists around each root in the rhizosphere. These organisms have formed a symbiotic relationship with plants by providing this protection in exchange for nutrients such as oxygen, carbohydrates, and inorganic minerals. In addition, certain soil microbes such as vesicular-arbuscular mycorrhizal (VAM) fungi also protect plants by restricting the uptake of non-essential inorganic elements such as Cd, Ni, Pb, etc. One hypothesis is that the VAM fungi sequester the metals into their tissue, restricting the transfer into the plant. Another hypothesis uses the enhanced water uptake capabilities provided by the fungi to dilute the concentrations of extracted elements in the plant root cells [13]. Although a clear understanding of the role of VAM fungi in the protection of plant roots to certain inorganics such as heavy metals is not available, their use warrants further research, particularly in a phytotechnology aspect. In addition to metal resistances, fungal organisms also provide plants with resistances to drought, salinity, and acidic soils in exchange for the carbohydrate nutrition [3]. Another beneficial aspect of soil hyphae is that they attract other microbes to form water-stable aggregates necessary for good soil tilth.

One aspect that is well known is the ability of VAM fungi to enhance the growth and nutrition of the plants themselves. Specifically, these fungi can extend out several orders of magnitude beyond the range of the rhizosphere (centimeters or meters instead of millimeters). However, the fungal symbiont maintains a direct connection to the plant root [3]. Fungal spores present in most soils germinate and penetrate into the cortical cells of plant roots growing in the vicinity [14]. The fungal hyphae then extend outwards through mycelial growth, seeking inorganic nutrients that can be traded with the plant for carbohydrates and other nutrients. Because these fungal hyphae extend out into the soil beyond the reaches of the root system itself, they can access regions of the soil beyond those that have been depleted of nutrients by the plants. Furthermore, the increase in total surface area provides additional water uptake capacities for the plant. It is in the arbuscular structures of the cortical cells where metabolites are exchanged between the fungal symbiont and the host cytoplasm. Furthermore, the vesicular structures formed in the cortical cells function as nutrient storage organs that can be utilized by the plant under nutrient stress conditions. Some results have shown nutrient uptake enhancements by the ectomycorrhizal fungi that are five times that of non-infected roots [15].

2.2 Plant Processes

Typical plant growth includes many different but interrelated processes. In the subsurface, plant roots take up water and absorb inorganic nutrients while exuding phytochemicals into the rhizosphere. Terrestrially, plant leaves gain energy through solar radiation to photosynthesize the phytochemicals from atmospheric CO_2. In the complex gas exchange processes occurring from the leaves, plants release oxygen and transpire water vapor back into the atmosphere [16]. Each of these biological process can contribute to the containment and/or remediation of contaminants.

2.2.1 *Inorganic Nutrition*

The 13 essential inorganic nutrient ions include nitrate or ammonium nitrogen (NO_3^- or NH_4^+), phosphate ($H_2PO_4^-$, HPO_4^{2-}, or PO_4^{3-}), potassium (K^+), calcium (Ca^{2+}), magnesium (Mg^{2+}), sulfate (SO_4^{2-}), iron (Fe^{2+} or Fe^{3+}), chloride (Cl^-), zinc (Zn^{2+}), manganese (Mn^{2+}), copper (Cu^+ or Cu^{2+}), borate (BO_3^{3-} or $B_4O_7^{2-}$), and molybdate MoO_4^{2-}) [17]. These mineral nutrients are taken up by the root system as dissolved constituents in the soil water. These elements are required by the plant for growth, development, or reproduction. The uptake of these nutrients can occur through the apoplasm (between root cells) or the symplasm (from cell-to-cell crossing root cell membranes) and are acquired either passively in the transpirational stream (see Sect. 2.2.2) or actively through transport proteins associated with the root membrane. Two common modes of mineral uptake are passive electrochemical gradient channels and ATPase (adenosine triphosphatase) mediated transport proteins. Once inside of the root system, the dissolved inorganic elements can be transported throughout the remainder of the plant through the vascular system of the plant known as the xylem (see Fig. 2).

In addition to the essential nutrients, other non-essential minerals, that are considered environmental concerns such as salts, As, Cd, Na, Se, and Pb, can be taken up as well. Several of these inorganic elements can be taken up through the

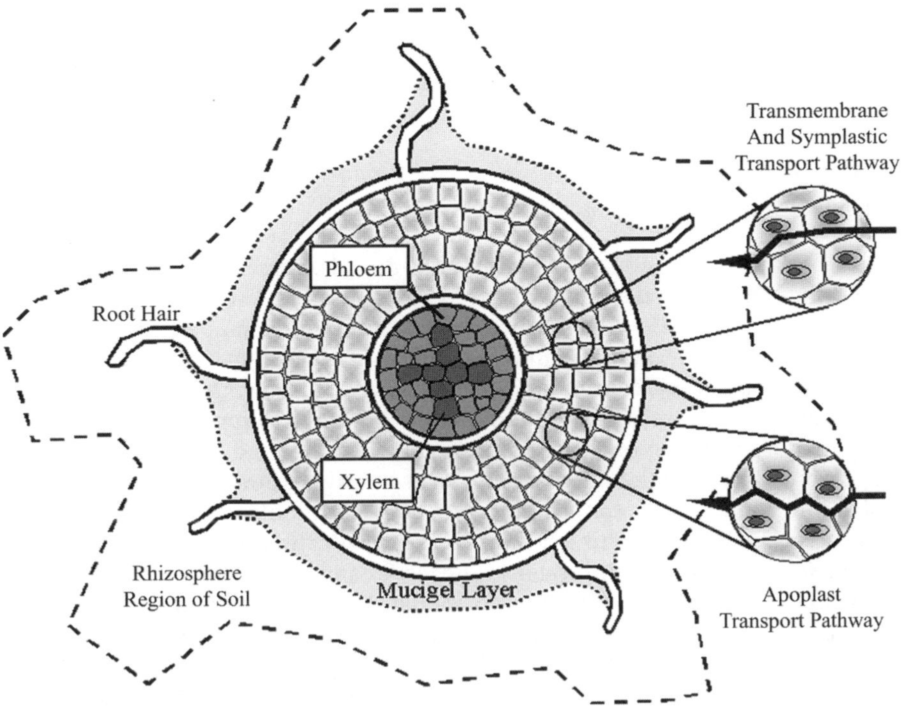

Fig. 2. Schematic cross-section of a primary root (upper bubble = transmembrane and symplastic transport pathways through the root cells, lower bubble = apoplastic transport pathway between root cells)

passive channels while others, typically multivalent cations, can substitute onto transport proteins and be taken up instead. Similarly, elements normally considered plant nutrients, such as Cu, Zn, and Mn, can also raise environmental concerns when present in high concentrations. Some of these minerals are considered regulated contaminants while others simply render a soil arid or infertile leading to erosion and instability.

Because these inorganics are not essential or can be toxic to the plant at high concentrations, the plant utilizes various mechanisms to sequester or stabilize these extraneous inorganics and prevent translocation into the more sensitive, terrestrial portion of the plant [18]. One primary mechanism is to bind these inorganics in the soil by exuding biochemicals that facilitate the precipitation or adsorption onto the soil matrix. Similarly, the inorganics can be irreversibly bound onto the root surface proteins themselves, preventing them from entering into the plant. Another mechanism is to sequester the non-essential inorganic into the vacuoles of the plant cells which act, in part, as a storage receptacle for the plant and help prevent further transport into the plant system. A final mechanism is to reduce the toxicity of the inorganic, but allow its uptake by changing its speciation through incorporation into an organometallic compound [19]. Depending on the fate of the inorganic in the soil-plant system, a suitable phytotechnology-based system can be developed for impacted sites. The primary mechanisms for cleaning up inorganics are sequestration in the rhizosphere (phytosequestration) or extraction and subsequent removal into the terrestrial plant tissues (phytoextraction).

2.2.2
Water Uptake and Transpiration

Subsurface nutrients and contaminants alike must be present in the soil aqueous phase in order for uptake into vegetation to be possible. The processes of water uptake and transpiration begin when soil moisture is taken up by the roots and end when the water evaporates into the atmosphere through the leaves. The water uptake phase of this continuum occurs primarily through the apoplastic pathway, although it is present throughout the cellular environment, including the transmembrane and symplastic pathways. At the other end, the evaporation of water occurs through the microscopic pores in leaves known as stomata. While inside the plant, the process of water transport from roots to shoots to leaves is known as translocation. This whole process occurs primarily by the equilibrium driving force between liquid water in the leaves and the gaseous water (humidity) in the atmosphere [16]. In order to regulate this process, plant roots produce abscisic acid into the xylem as a biochemical signal to the stomata to alter their aperture size [20]. As plants become stressed, the size of the stomatal aperture is reduced to decrease the water vapor exiting from the plant. The stomatal conductance (the tendency for water vapor to diffuse out of the stomata) can be monitored using a leaf porometer or pressure bomb.

In terms of phytotechnologies, this process of water uptake from the subsurface and transpiration into the atmosphere is the primary mechanism used to control hydraulics and contain contaminants. Furthermore, certain contaminants that are readily soluble in water can be taken up and even transpired through the leaves. This area of phytotechnology (phytovolatilization) has been useful for several common contaminants such as volatile organic compounds (VOC) and certain metalloids.

2.2.3
Photosynthetic Production of Phytochemicals

The carbon dioxide present in the atmosphere enters into plants through stomata in a counter current gas exchange with water vapor. The carbon dioxide is then incorporated into photosynthetic products (phytochemicals) using biochemical reductants generated during photosynthesis. The two primary biochemical pathways that exist within plants for incorporating this carbon into phytochemicals are the C3 and C4 carbon pathways. The C3 pathway refers to the 3-carbon compound, 3-phosphoglycerate, while the C4 pathway refers to the 4-carbon compounds, malate or aspartate. These compounds represent the first stable compounds where carbon dioxide is sequestered in the plant. After incorporation into these compounds, the carbon is eventually processed in the photosynthetic carbon reduction cycle which leads to a variety of compounds including cellulose, lignin, starch, carbohydrates, sugars, amino acids, proteins, enzymes, and other cellular materials.

The products of these reactions, known as photosynthates, are translocated throughout the plant, even down into the root system. This occurs through another vascular system known as the phloem (counter-flow to the xylem). These products can be incorporated into the biomass, metabolized during cell respiration to produce energy, or exuded into the root zone.

2.2.4 Root Exudation and Root Turnover

Some estimates show that the amount of carbonaceous material exuded by plants can be as high as 50% of the total photosynthetic production [21]. One of the major forms of root exudation material is mucigel, the secretions from the growing regions around roots that aids in root penetration, protects the root apex from desiccation, and promotes nutrient uptake [16]. The composition of this organic material in the rhizosphere varies with plant species; therefore, the primary colonizers for specific plants vary considerably as well.

In addition to mucigel, plants also secrete a broad range of metabolites as a mode of defense from potential soil pathogens such as harmful viruses, bacteria, fungi, insects, nematodes, etc. Several of the common types of secondary plant defense compounds (phytoalexins) include terpenoids, flavanoids, alkaloids, phenylpropanoids, anthraquinones, polyphenolics, and others [22, 23]. These are derived from the three basic secondary metabolic pathways: acetate-malonate, acetate-mevalonate, and shikimate pathways. Whether these phytoalexins are actively exported from the roots or are released due to the senescence (death) of the root tissue varies widely. This is an area under current investigation.

Similarly, plants also conduct a form of chemical warfare against other plant species in their vicinity by producing chemicals which prevent their growth. This process is termed allelopathy and ensures that competition for nutrients and water are minimized. Perhaps the best known example of allelopathy is the production and exudation of juglone from eastern black walnut trees (*Juglans nigra*) [24].

In addition to the active exudation of phytochemicals, most plants also experience an annual cycle where a portion of the root system is sloughed off by the plant prior to dormancy in the winter months. This is termed root turnover and can lead to a substantial deposition of additional organic material that is made available to soil organisms to feed upon [25]. Recent estimates have shown that 58% of the fine roots of mulberry trees (*Morus rubra*) died in autumn leading to a 2-fold increase in phenolic deposition in the soils [26]. This influx of additional carbon helps to maintain the bioactivity of the soil organisms during the winter months. In addition to a source of carbon, the extricated plant material also caused the release of VAM fungal hyphae which are freed to inoculate additional plants during subsequent growing seasons [14].

In terms of phytotechnologies, the production and exudation of organic compounds and the turnover of plant root systems represents a major source of carbon for soil organisms. By sustaining various communities of soil organisms in the rhizosphere, plants are capable of growing in soils and utilizing moisture that is otherwise impacted with contaminants. This area of phytotechnologies (rhizodegradation) represents one of the primary methods for cleaning up organically impacted sites. There are even successful demonstrations for relatively recalcitrant contaminants such as pentachlorophenol (PCP), polychlorinated biphenyls (PCB), polycyclic aromatic hydrocarbons (PAH), and other semivolatile organic compounds (SVOC) [25, 27–30].

2.2.5
Bio-Metabolism

Although the rhizosphere soil organisms represent a major resource for reducing organic contaminants in the subsurface, plants are also capable of producing proteins, enzymes, and cofactors that have bioactivity for certain contaminants. A whole host of plant-derived enzymes have been investigated including nitroreductases, dehalogenases, glutathione, oxygenases, phenoloxidases, nitrilases, and phosphatases. Each of these enzymes have been reported as responsible for the breakdown a broad array of contaminants in solid and liquid media. Nitroreductase enzymes are able to breakdown nitroaromatics such as the explosive compounds: trinitrotoluene (TNT), hexahydro-1,3,5-trinitro-1,3,5-triazine (RDX), and octahydro-1,3,5,7-tetranitro-1,3,5,7-tetraazocine (HMX) [31]. Dehalogenases are able to cleave the halogen components off of contaminants such as chlorinated solvents. Specifically, perchloroethylene (PCE), TCE, and dichloroethylene (DCE) can be eventually reduced to vinyl chloride (VC) in the reductive dechlorination process catalyzed by dehalogenases [32]. Alternatively, other plant-based enzymes such as glutathione and (cytochrome P450 mono-) oxygenases have been shown to be successful at oxidizing these contaminants. Specifically, the oxidation of TCE yielded trichloroethane (TCA) and trichloroacetic acid (TCAA) as stable products [33, 34]. Another group of plant enzymes known as the phenoloxidases (laccases, peroxidases, tyrosinases) have also been shown to be effective on the chlorinated solvents as well as a broad range of phenolic compounds [35]. Finally, nitrilases and phosphatases have been discovered from plants that can breakdown various pesticides (herbicides, insecticides) [36]. The use of these plant-derived enzymes represents another one of the major areas of phytotechnologies (phytodegradation) used to clean up organic contaminants. Furthermore, phytodegradation is closely related to rhizodegradation in that several of these enzymes are exported or released during root turnover.

3
Mechanisms

The basic physiological processes occurring in the root-zone, within the vegetation, and through the transpiration stream are utilized in phytotechnologies as the foundations for designing the final systems. In each region of this soil-plant-atmosphere continuum, various processes are available to provide clean up, but depend on the type of contaminant present. The ability of a contaminant to pass through the various regions of this continuum depends on its physical characteristics such as solubility (C_w^{sat}), Henry's Law gas constant (K_H), vapor pressure (P_0), and phase distribution or partitioning coefficients (K_d). These parameters are available in several references [3, 37–39].

3.1 Occurring Outside Plant System

In the subsurface region of the continuum, inorganic contaminants are subject to various soil organisms, plant enzymes, and phytochemicals that can lead to the sequestration in the rhizosphere. In phytotechnologies, these basic physiological processes are collectively known as the mechanism of phytosequestration. For organic constituents, the rhizosphere also provides several processes for breaking down the contaminants. The direct and co-metabolic biodegradation carried out by the soil organisms associated with the plant rhizosphere is termed in phytotechnologies as rhizodegradation.

3.1.1 *Phytosequestration (Inorganics)*

Phytosequestration is the precipitation or immobilization of inorganic contaminants in the soil, on the root surfaces, or within the root tissues. The ability of plants to sequester contaminants in the root zone is a form of protection to prevent the translocation of the potential toxin to the more sensitive terrestrial tissues such as the reproductive organs. There are various steps along the soil-plant-atmosphere continuum where inorganics can be sequestered and prevented from migrating in the environment. The relative distances in the soil-root environment are shown in Fig. 3. In the surrounding soil environment, organisms such as VAM fungi can actively sequester the minerals into their biomass. This incorporation prevents further migration into the plant. The presence of sulfates, hydroxides, oxides, carbonates, and carboxylates deposited in the rhizosphere by plants can lead to the precipitation of certain elements into the soil matrix [40]. Furthermore, changes in soil pH brought about by root exudates such as organic acids and bases can facilitate the precipitation of various ionic species [36]. Similarly, the redox potential of the soil can be altered by various microorganism or plant-derived chemicals leading to a conversion in electron state of the inorganic. Changes in the oxidation state can alter the solubility [41]. Adsorption of inorganic elements onto roots at the cell wall occurs through ion exchange, complexation to carboxyl ligands, and electrostatic binding [42, 43]. Once inside of the root tissues, the minerals can be stored into the root vacuoles or in the apoplastic pool [40].

Some of the inorganic contaminants which have been addressed using phytosequestration include Au, Cr, Cu, Fe, Pb, and Zn [36, 40, 41, 43, 44]. This is not meant to be an exhaustive list as additional inorganic contaminants susceptible to the phytosequestration mechanisms are areas of active research and development. As contaminants accumulate onto or into the root tissues, the efficiency of adsorption can be measured using the Root Concentration Factor (RCF) [2]. The

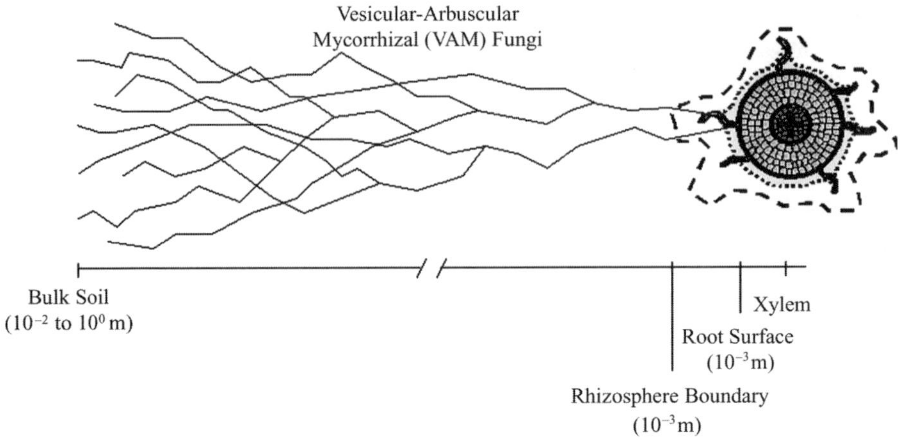

Fig. 3. Relative distances in the soil-root region of the continuum where phytosequestration of inorganics occurs

RCF is defined as the ratio of the contaminant in or on the roots to the concentration dissolved in soil water (g/kg root per g/L). More efficient phytosequestration plants contain higher RCF values, typically greater than unity. Note that the RCF is also applicable for certain organic contaminants.

3.1.2
Rhizodegradation (Organics)

Rhizodegradation is the breakdown of organic contaminants by subsurface organisms that are supported or enhanced, in some manner, by the presence of plants. In general, the rate limiting factor in the biodegradation of organic contaminants in the subsurface is the concentration of dissolved oxygen (or electron acceptors, in general). Plants, being aerobic organisms, maintain an oxygenated environment in the root zone through one of two primary mechanisms. The first is that plants, particularly wetland species, are capable of producing and exporting oxygen into a saturated medium. Estimates on the rate of oxygenation are in the range of 0.5 mol O_2 per m^2 surface area per day [1]. Secondly, the channels created by roots, particularly after root turnover, allow atmospheric oxygen to diffuse into the subsurface [36]. These two sources of oxygen sustain aerobes growing in the rhizosphere and enhance the biodegradation of organic contaminants. Furthermore, the release of phytochemicals either through exudation or root turnover provides additional food sources for the microbial community. These microbes, in turn, biodegrade contaminants either directly through metabolism as a carbon source or indirectly through co-metabolic biodegradation. In addition to releasing phytochemicals, the plant roots also provide various enzymes that aid in the breakdown of the contaminants. Working together, the plants and soil organisms are able to stimulate the effective clean up of the subsurface.

The symbiotic relationship between plants and soil microbes has proven to be beneficial to both classes of organisms. Soil microbes receive various nutrients such as oxygen, inorganic nutrients, carbon sources, and water. These can be derived directly from the plant root systems themselves through phytochemical exudation, root turnover, plant-produced enzymes, and the transpirational stream. In return, plants receive enhanced water uptake, additional nutrient sources, as well as protection from potential pathogens and various anthropogenic contaminants. Plants produce extensive root systems that support a wider diversity of soil organisms. This increased community structure is an indicator of a healthier soil condition that can eventually be returned to a cleansed state.

Since contact between the root system and the impacted media is paramount to the success of this phytotechnology mechanism, then the extensiveness (density and depth) of the root system plays an important role in treating organically impacted sites. Typically, turf grasses such as bluegrass (*Poa* sp.) have very shallow but dense, fibrous root systems that only extend down a few inches [45]. Other grasses such as ryegrass (*Lolium* sp.) and fescue (*Festuca* sp.) as well as some herbaceous species such as clovers (*Trifolium* sp.) and various vetches (*Astragalus*, *Coronilla*, *Lotus*, and *Vicia* sp.) contain root systems that typically can penetrate down on average between 1 to 4 feet below ground surface (bgs) [25, 36]. Likewise, many woody species are typically shallow rooted, requiring lateral root growth to maintain support. On the other hand, some prairie grasses and wildflowers can produce root systems that are 10 to 15 feet deep while maintaining a reasonable density for creating larger rhizosphere regions [29, 45]. Several of these species are shown in Fig. 4. Similarly, certain trees known as phreatophytes are capable of tapping directly into the saturated zone. The root systems of representative trees such as poplars (*Populus* sp.), willows (*Salix* sp.), and eucalyptus (*Eucalyptus* sp.) typically attain depths of 10 to 15 feet bgs although reports of deeper root systems are often reported [1, 36, 46]. Other herbaceous species such as alfalfa and certain medicks (*Medicago* sp.) can develop deep tap root systems that can penetrate several meters bgs. However, the density of the root systems fall dramatically with depth with roughly 70% to 90% of the total root mass remaining in the surface soils [1].

Several different organic contaminants have been remediated through this mechanism. Substantially reduced concentrations have been exhibited when measuring either individual components such as BTEX and PAH [25, 29, 47–49] as well as in broader hydrocarbon fractions such as gasoline range organics (GRO), diesel range organics (DRO), oil and grease (O&G), and total petroleum hydrocarbons (TPH) [30, 50, 51]. Other organic contaminants include pesticides such atrazine, metolachlor, triflurain, parathion, diazinon, propanil, and 2,4-dichlorophenoxyacetic acid (2,4-D) [36, 52] and chlorinated compounds such as TCE, PCP, and PCBs [27, 28, 53]. The rates of biodegradation are typically in the range of published aerobic biodegradation rates. However, in comparison to mechanical agitation for oxygenating the soils, concentration reduction rates were roughly doubled using rhizodegradation [51].

A) Kentucky Bluegrass (*Poa pratense)*
B) Lead Plant (*Amorpha canescens*)
C) Missouri Goldenrod (*Solidago missouriensis*)
D) Indian Grass (*Sorghastrum nutans*)
E) Compass Plant (*Silphium laciniatum*)
F) Porcupine Grass (*Stipa spartea*)
G) Heath Aster (*Aster ericoides*)
H) Prairie Cordgrass (*Spartina pectinata)*
I) Big Blue Stem (*Andropogon gerardii*)
J) Pale Purple Coneflower (*Echinacea pallida*)
K) Prairie Dropseed (*Sporobolus heterolepis*)
L) Side Oats Grama (*Bouteloua curtipendula*)
M) False Boneset (*Kuhnia eupatorioides*)
N) Switch Grass (*Panicum virgatum*)
O) White Wild Indigo (*Baptista leucantha)*
P) Little Blue Stem (*Andropogon scoparius*)
Q) Rosin Weed (*Silphium perfoliatum*)
R) Purple Prairie Clover (*Petalostemum purpureum*)
S) June Grass (*Koeleria cristata*)
T) Cylindric Blazingstar (*Liatris cylindracea*)
U) Buffalo Grass (*Buchloe dactyloides*)

Fig. 4. Diagram of the extensiveness of various root systems (adapted from reference [45])

3.2
Occurring Inside Plant System

If a contaminant is able to pass through the rhizosphere, it may then be subject to various processes occurring within the vegetation. For inorganic contaminants, those minerals that are not sequestered in the root zone can be taken up and accumulated into the terrestrial portion of the plants. Subsequent harvest of the terrestrial tissues removes the contaminant from the site. This mechanism is known in phytotechnologies as phytoextraction. Similarly, certain organic contaminants or by-products of rhizodegradation can pass through the rhizosphere. However, these are then subject to degradation by enzyme present within the plants themselves. This phytotechnology mechanism is known as phytodegradation.

3.2.1
Phytoextraction (Inorganics)

Phytoextraction is the removal of inorganic contaminants from the subsurface through plant uptake. This includes heavy metals, metalloids, radionuclides, and salts. The two special classes of plant species that have been investigated for their uptake capabilities to a wide range of inorganic minerals are hyperaccumulators and halophytes. The definition of a hyperaccumulator is one that can concentrate a specific metal to at least 1000 mg/kg (0.1% dry weight of tissue) or 10000 mg/kg (1%), depending on the specific inorganic [54]. These concentrations are generally several times normal concentrations that are typically regarded as toxic to most other plant species. Similarly, halophytes are species that can tolerate or accumulate salinity levels present in the form of sodium, calcium, or magnesium chloride, or as free chloride ions between 3 and 7 times normal concentrations (less than 10000 parts per million, ppm, or 1% total salts) [55]. Not all halophytes are accumulators. Some species simply tolerate high saline conditions while others actually remove salt from the subsurface and move it into the terrestrial tissues. Phytoextraction involves the latter.

Some of the inorganic constituents which have been shown to be extracted by plants are certain metals (Ag, Au, Cd, Co, Cr, Cu, Hg, Mo, Ni, Pb, Tl, V, Zn) [36, 41, 44, 56–61], metalloids (As, Se) [62, 63], radionuclide elements (^{137}Cs, ^{3}H, ^{239}Pu, ^{90}Sr, $^{234/238}$U) [36, 64–66], and the aforementioned salts. Some of the well known plant genuses that contain species of hyperaccumulators or halophytes include *Atriplex*, *Brassica*, *Helianthus*, *Kochia*, *Pelargonium*, *Pinus*, *Salicornia*, and *Thlaspi*.

The uptake of these constituents can arise during nutrient deficiency conditions which is often the case at environmentally impacted sites [67]. One of the primary mechanisms used by hyperaccumulators to take up and store high concentrations of these inorganic elements is to produce exudates that react with the constituent to form a less toxic organometallic compound. These organometallics can then be taken up safely and stored in the cell vacuoles [19]. For halophytes, two primary mechanisms are utilized. The first is the ability to preferentially take up additional water (via VAM fungi) in order to dilute the saline concentrations within the tissues [13]. Secondly, certain plants such as salt cedar (*Tamarisk* sp.) and salt bush (*Atriplex* sp.) contain specialized organs called salt glands that can store and regulate the cellular osmotic potential [16].

3.2.2
Phytodegradation (Organics)

Phytodegradation is the ability of plants to produce the enzymes or cofactors necessary to degrade organic contaminants. In some cases, these catalysts may be exuded; therefore, phytodegradation is highly related to rhizodegradation. An area of current research is whether specific plant-based biomolecules are active inside, outside, or in both regimes of the soil-plant environment. Specific classes of chemicals which have been investigated include petroleum hydrocarbons, chlorinated compounds, pesticides, munitions, and phenolics.

In order for an organic chemical to enter into a plant, it must contain the proper chemical properties to transfer through the root membrane. These properties include solubility, polarity, hydrophobicity, and partitioning coefficients (K_d). In a series of tests conducted on a number of organic contaminants, the octanol-water partition coefficient (log K_{ow}) was found to be a key factor describing the potential for uptake [2]. Many organic chemicals with a log K_{ow} in the range of 1 to 3.5 have the correct chemical properties to pass into the plant. Hydrophobic chemicals (log $K_{ow} > 3.5$) are generally not sufficiently soluble in water or are bound so strongly to the surface of the roots that they cannot be easily translocated into the plant. On the other hand, chemicals that are highly polar and very water soluble (log $K_{ow} < 1.0$) are not sufficiently sorbed by the roots nor are they actively transported through plant membranes due to their high polarity [68].

Similar in concept to the RCF, the transpiration stream concentration factor (TSCF) is determined by taking the ratio of the concentrations in the xylem sap and in the soil water (dimensionless) [2]. Unlike the RCF, most contaminants contain a TSCF value that is less than unity. This is due to the fact that vegetation will preferentially take up water while retarding the translocation of the potentially toxic chemical. Alternatively, the concentration of organic contaminants in the transpirational water is reduced due to the bio-metabolism occurring within the plant itself. The TSCF values do not include contaminants that have been sequestered in the plant tissues themselves, only the freely mobile portion in the transpirational stream. Note that the TSCF is also applicable to inorganics but is dependent on chemical speciation and redox state [2].

One area that is relatively unclear is whether a specific contaminant is taken up in its parent form or whether it is partially rhizodegraded and then subsequently taken up into the plant. This is further complicated due to the current restrictions in analytical techniques for distinguishing between contaminant compounds or by-products with the organic biomass of the plant tissues. Specifically, the concentrations of the chemicals are generally low, often in the same order of magnitude as the detection limits of the analytical procedure [69]. Furthermore, matrix interferences make specific contaminant peaks indistinguishable from the background noise on gas chromatograms. The use of mass spectroscopy has aided in identifying specific chemicals or by-products; however, the complete phytodegradation pathways for most organic contaminants remains undetermined. The use of radiolabeled compounds has been used to track the metabolic degradation of the contaminants [48, 70]. These have shown various fates for the organic compounds including incorporation into the plant lignin or the complete breakdown to CO_2 and water [2].

3.3
Occurring Through the Plant System

The conduit through which contaminants are able to pass from the soil and into the plant is through the transpirational stream. Those constituents which are soluble and volatile enough may even pass completely through the plant in the transpiration gases. In addition, the ability to control hydraulics through evapotranspiration has been used extensively as a phytotechnology mechanism.

3.3.1
Phytovolatilization (Inorganics and Organics)

Certain contaminants, inorganic and organic, are capable of being volatilized into the atmosphere while present in a dissolved form in the transpirational stream. This mechanism is known as phytovolatilization and depends heavily on the physical characteristics of the contaminant itself. In order to get into the plant, the contaminant should have the proper chemistry to pass through the root membrane (see previous discussions on phytoextraction and phytodegradation). Once inside the plant, organic chemicals that are generally considered relatively volatile such as BTEX, TCE, and MTBE, can be phytovolatilized into the atmosphere through the stomata [69, 71, 72]. Alternatively, a metabolic by-product may be more conducive to phytovolatilization than its parent compound and can be remediated using this mechanism. Physical properties that typically dictate the relative chemical volatility include the vapor pressure and Henry's Law gas constant. Once in the atmosphere, these chemicals are rapidly photodegraded by the hydroxyl radicals produced in the photochemical reduction of ozone. Typical photo-oxidation rates in air for these VOCs are 1 to 2 orders of magnitude faster than the degradation rates in soils or water [9].

In addition to volatile organics, certain metalloid inorganics such as Hg and Se are susceptible to this phytotechnology mechanism as well. In the case of mercury, genetically engineered *Arabidopsis thaliana* and tobacco plants (*Nicotinia tabacum*) expressing the merA and merB genes isolated from soil bacteria have been developed which can take up dissolved methylmercury, convert it to Hg(II), and then volatilize it as elemental mercury [73]. The environmental advantage of this process is the roughly six order of magnitude reduction in toxicity from methylmercury to elemental mercury. However, the introduction of volatilized Hg into the atmosphere through phytovolatilization has yet to be accepted in a commercial or regulatory application. Similarly for Se, this inorganic is transformed into a dimethylated organometallic which can be volatilized from the plant leaves. The dimethyl selenite is a less toxic form than the Se present as selenate [74].

3.3.2
Evapotranspiration

Evapotranspiration (ET) is the combination of water evaporation derived from the direct exposure of water to the atmosphere and the transpiration of soil moisture by plants. Each of the five mechanisms defined above, phytosequestration, rhizodegradation, phytoextraction, phytodegradation, and phytovolatilization, are mechanisms which can lead to the treatment of specific contaminants. This last mechanism of phytotechnologies is the primary means of containing contaminants by controlling hydraulics. However, these mechanisms are interrelated as the transpirational uptake of soil moisture is necessary in order to bring the dissolved contaminants into the soil-plant-atmosphere continuum.

Soil moisture is derived from many sources, primarily from precipitation. The initial contact that plants have with water is through the direct interception of rain by the terrestrial portion of the plants, typically the leaves. The rain interception capacities of various plant and trees have been determined primarily for irrigation requirements in agriculture, forestry, landscaping, and landfill cover designs. Typical rain interception capacities for a natural pasture or undefined mixture of species is 14% to 26% [75]. These values are the same order of magnitude for a forest canopy created as a single species monoculture (10% to 30% depending on species). However, certain plant species such as native prairie grasses are able to intercept greater quantities of rain and prevent this water from reaching ground surface. Using a rainfall rate of a half inch of rain in 30 minutes, these native grasses were capable of intercepting between 50% and 60% of the water [75]. Because this water is intercepted on the leaf surfaces and can subsequently be evaporated after the rainfall has ceased, this mechanism provides a means to reduce infiltration and groundwater recharge. This ability to control hydraulics from the surface represents the evaporative component of the evapotranspiration mechanism.

The physiological factors affecting the ability of vegetation to intercept rain include leaf structure (narrow blades versus wide leaves), morphology (waxy versus hairy cuticle), orientation (vertical versus horizontal positioning), and canopy development (layers of leafy structures). Typical prairie grasses contain narrow, hairy leaf blades positioned vertically in a dense, multi-layered canopy. The crevices created by the vertical leaves with the stalk provide catchment points for significant volumes of water.

If the rainfall event is able to overcome the rain interception capacity of a canopy, then this water is able to infiltrate into the soil. However, this water is then subject to the transpiration process of water uptake by the roots and release into the atmosphere through the plant leaves. That water which is able to pass through the root zone is then capable of recharging groundwater. This, of course, depends on the rooting depth of the vegetation.

Transpiration rates are highly dependent on the plant species, local climate, soil conditions, and age of the canopy. One school of thought is that the transpiration rates can only achieve a fraction (typical 0.7 to 0.8 times) the maximum as determined by pan evaporation rates [76, 77]. Pan evaporation is the rate of evaporation from a free surface of water, typically measured using a (Class A) pan of known surface area. However, others believe that these rates can be exceeded due to the increased evaporative surface area provided by multiple layers of leaves [76]. Direct measurements of plant transpiration rates can be conducted using the heat pulse method [78]. Basically, this method utilizes a heat source in combination with thermocouples that measure the dissipation of heat due to sap flow through the xylem. Using empirically derived equations such as the Penman-Monteith equation, sap flow rates are converted into plant transpiration rates, taking into account general climatology data such as the net solar radiation, relative humidity, air temperature, and wind speed [79]. Furthermore, this estimate of potential evapotranspiration (PET) includes other site-specific characteristics such as soil temperature, canopy structure, and leaf characteristics.

$$PET = \frac{\Delta(R_n - G) + \rho\, C_p\, D\, g_a}{\lambda[\Delta + \gamma(1 + g_a/g_c)]} \tag{3}$$

In Eq. (3), Δ is the rate of change of saturation vapor pressure (Pa/°C), R_n is the net solar radiative flux density (in W/m^2), G is the soil heat flux density (in W/m^2), ρ is the air density (kg/m^3), C_p is the specific heat of air (J/kg°C), D is the vapor pressure deficit (Pa), g_a is the aerodynamic conductance (m/s), g_c is the canopy conductance (m/s), λ is the latent heat of vaporization (J/kg), and γ is the psychrometric constant (Pa/°C).

These measurements are often confounded by several factors including wind, micro-climates in different portions of a canopy, and the positioning of the sap flow sensors on the vegetation [80]. Furthermore, the arguments for maximum versus exceedance of pan evaporation rates is confounded when taking into account the availability of groundwater to certain phreatophytic species, representing excess sources of water. Typical transpirations for a variety of species types are presented in Fig. 5 and include many common species used in phytotechnology applications [33, 46, 79, 81–101].

Although the maximum or exceedance of the pan evaporation rate is an ongoing discussion, a known maximum that cannot be exceeded occurs when the canopy has reached closure. This occurs when the ratio of leaf area to ground area has achieved a value that typically ranges significantly above unity, but is highly dependent on species. The leaf area index (LAI) is a measure of this ratio and varies according to the age of the canopy and seasons. After canopy closure has been achieved, the addition of more leaves does not result in higher transpiration rates since shading decreases the water loss by decreasing the net radiation captured by the leaf (see Eq. (3)).

In order to measure the effectiveness of a vegetative canopy to intercept rain and evapotranspire the water, several techniques and measurement devices are available. Popular methods include soil moisture probes that measure the electrical conductivity of the soil, lysimeters that collect water that passes through the root zone, and tensiometers that measure soil matric potentials related to soil water content. Other methods include gamma densitometry, neutron moderation, resistance blocks, and time-domain reflectometry [102, 103].

4 Applications

The specific mechanisms that are exploited depend on several factors including the specific contaminant, current site conditions, the remedial objectives, and the regulatory issues. Containment strategies are designed to reduce the overall risks to human health and the environment by preventing the migration or transfer of contaminants to potential receptors. Treatment strategies, on the other hand, are designed to reduce the risks to human health and the environment by reducing the concentration and/or mass of toxic substances. In phytotechnolo-

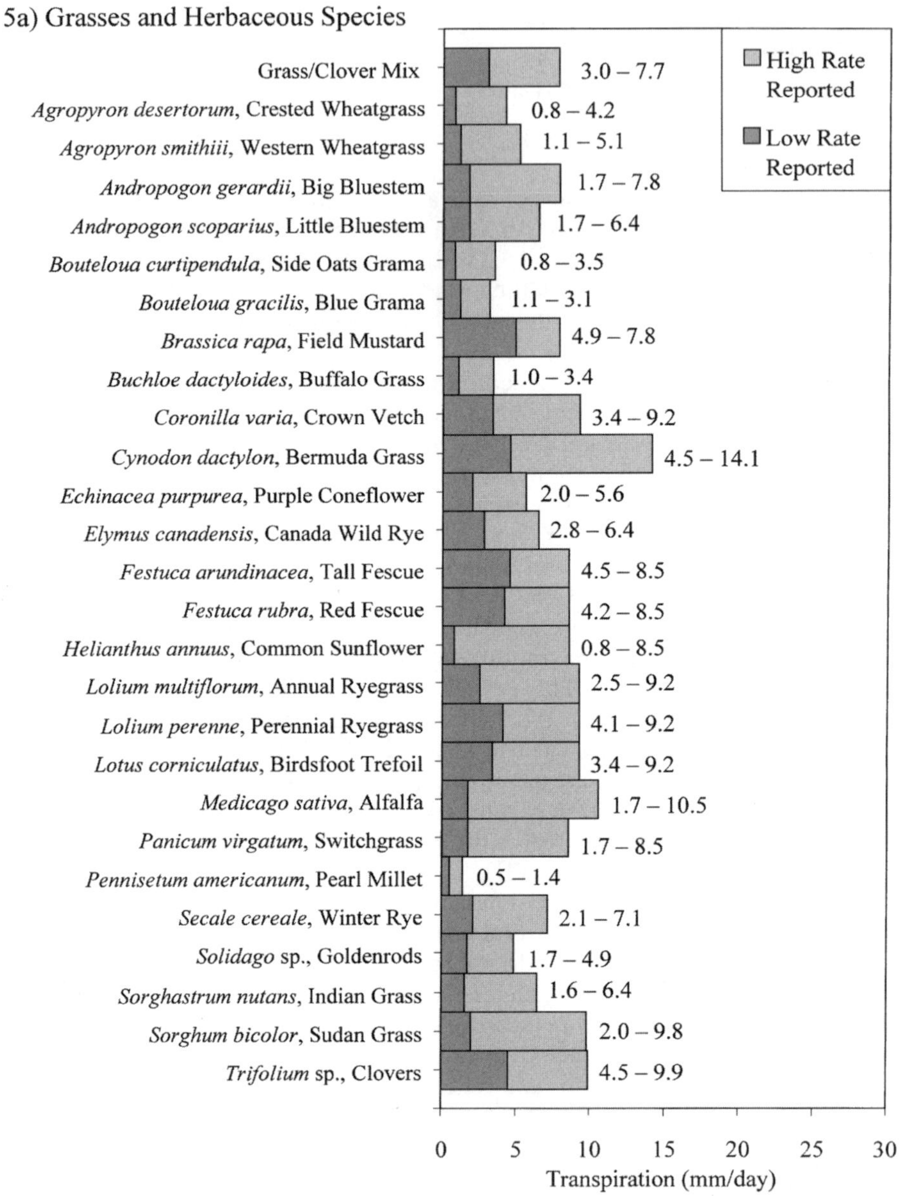

Fig. 5. Transpiration rates for a variety of plant species types. **5a** grasses and herbaceous species in mm/day, **5b** herbaceous wetland species in mm/day, **5c** woody species in gpd per tree

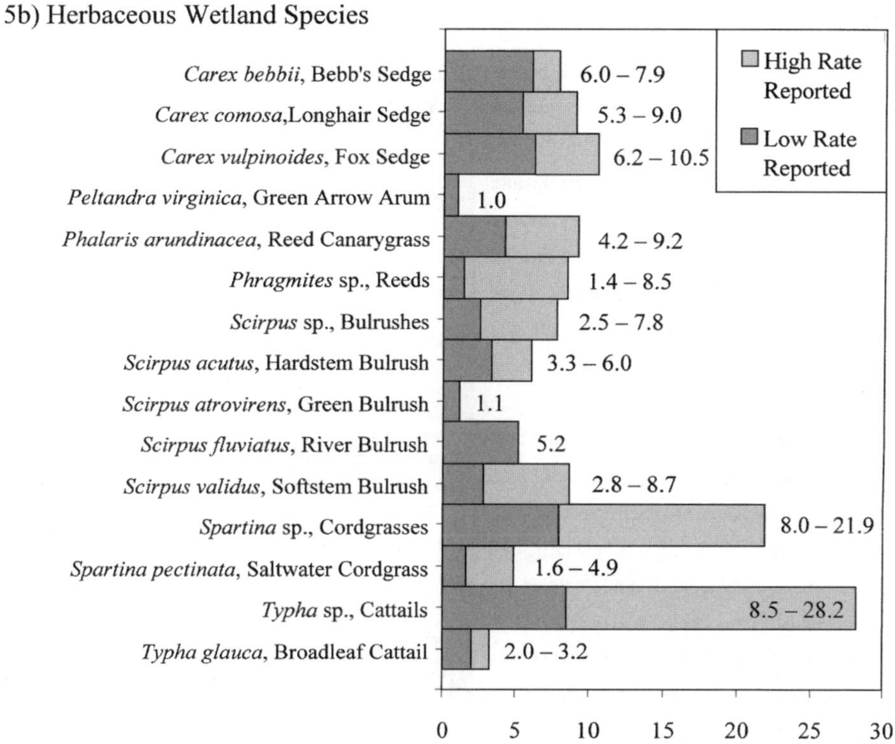

Fig. 5 (continued)

gies, certain applications utilize multiple mechanisms to achieve either or both of these goals.

4.1
Containment Strategies

The primary mechanism used in most containment strategies involves the plant evapotranspiration processes. This includes the control of groundwater using deep rooted species as well as the interception and subsequent transpiration of rain water to control surface infiltration. However, the use of phytosequestration as a means to reduce the mobility of various contaminants (primarily inorganic constituents) has also been applied as a containment system.

4.1.1
Groundwater Hydraulic Barriers

Deep-rooted species, particularly trees, are often used in phytotechnology applications to provide hydraulic control and plume containment through the evapotranspiration mechanism. The significant volumes of water that can be processed through a stand of trees or a field of high-transpiring prairie grasses

5c) Woody Species

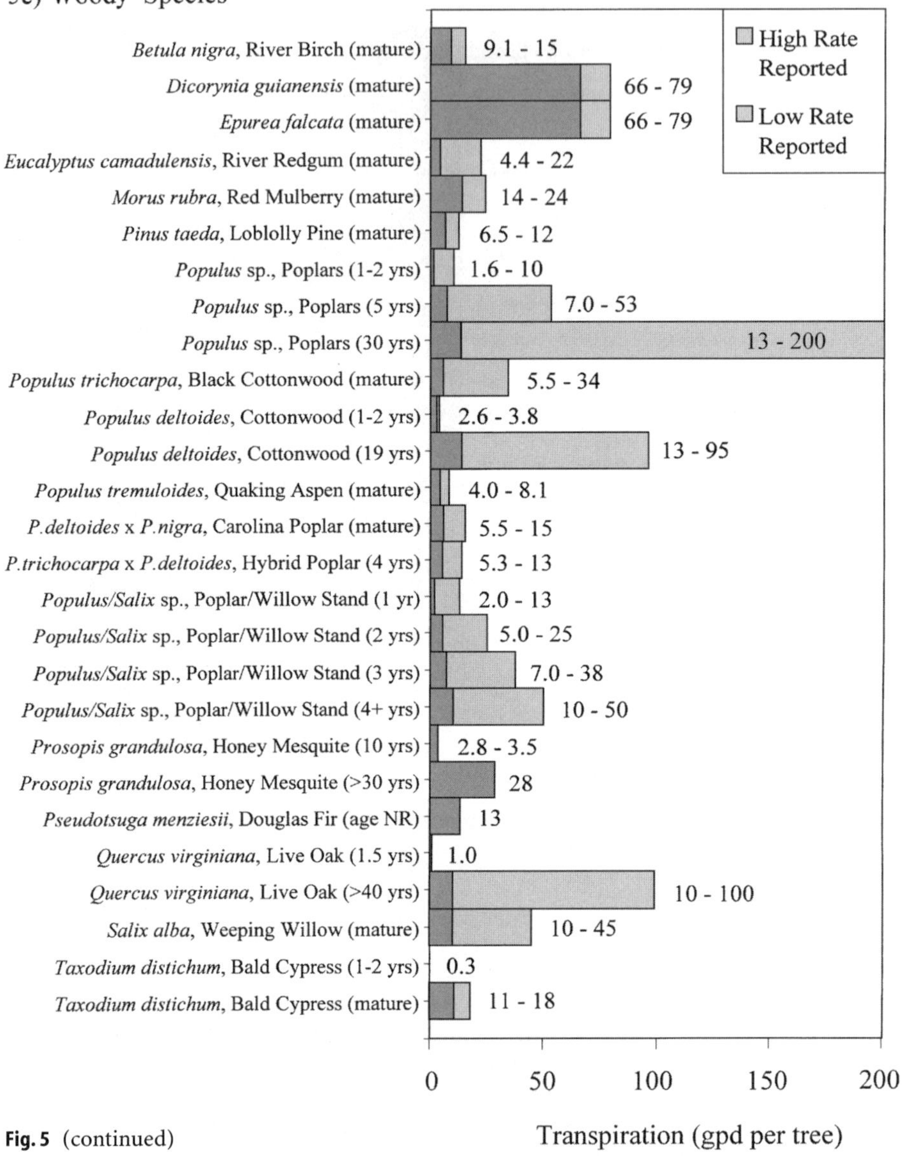

Fig. 5 (continued)

(see Fig. 5) can create depressions in the local water table which can be used to prevent plume migration. This local draw-down in the water table creates a hydraulic gradient which blocks the movement of groundwater beyond the planted area. This is shown schematically in Fig. 6.

In general, the planted area is situated perpendicular to the groundwater flow direction. Typical applications include the prevention of clean groundwater from entering into a contaminated zone or preventing contaminated groundwater from leaving a specified area. The former application is relatively easy to implement as contaminant toxicities to the vegetation are eliminated as potential impediments since the plantation is situated upgradient of the plume. The latter application is more commonly applied where the plantation is located downgradient of the contaminated area. Therefore, contaminated groundwater flows into the vegetated area and can subsequently be take up through transpiration in order to provide plume containment. Since the contaminants are brought into the root-zone, other phytotechnology mechanisms (phytosequestration, rhizodegradation, etc.) are then able to treat the contaminants and clean up the groundwater (see Section 4.3.2).

Several factors must be considered when designing these hydraulic control systems. In the simplest sense, the hydraulic containment system must be able to take up and transpire all of the water entering into the zone of capture. Therefore, the volumetric input of groundwater into the vegetated area must be estimated. This volume is dictated by the linear groundwater flow rate, also known as the Darcy velocity (v_D), and the width and thickness of the plume (W and D_t, respectively). The volumetric groundwater flow rate (Q), calculated in Eq. (4), must be less than or equal to the rate of uptake (transpiration) by the vegetation (V_T). However, one consideration that should be included for the movement of groundwater contaminants, particularly organic compounds, is that retardation due to physical-chemical processes such as adsorption onto soil colloids will reduce their actual velocity. The resulting seepage velocity (v_C) for the contaminants can be estimated as the Darcy velocity multiplied by the compound-specific retardation factor ($f_R < 1$) under the geological conditions at the site. In Eq. (4), v_C, describing the flow of contaminants, can be substituted for v_D, describing the flow of groundwater.

$$Q = (W\, D_t\, v_D) < V_T \tag{4}$$

$$v_C = v_D \cdot f_R \quad \text{where} \quad f_R < 1 \tag{5}$$

In addition, when designing such systems, the local rainfall (less leaf intercepted water – see Section 3.3.2) needs to be accounted for since vegetation will preferentially use this more readily available water instead of groundwater and produce root systems accordingly. The rate of supply of rainwater (R′) falling onto the area can be calculated using the area of the plantation (A) multiplied by the magnitude of the rainfall event (R, typically reported in inches or mm of water). Basically, the ground surface of the plantation will receive a certain fraction of this rainwater (f) depending on run-off contours (r) and canopy interception capacities (i), both reported as fractions (or percentages) of R. Since run-off from adjacent areas may drain into the planted area, the plantation may receive more

6a) Upgradient of Plume

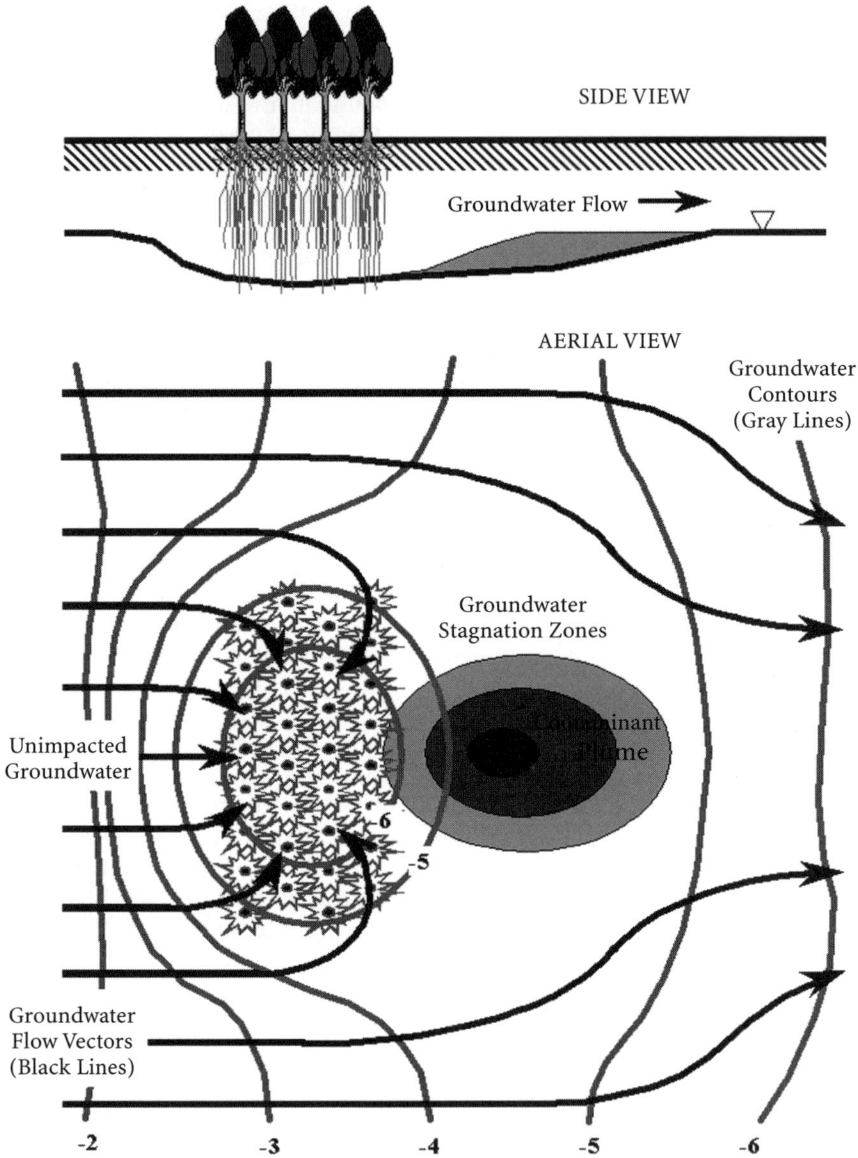

Fig. 6. Theoretical draw-down and groundwater flow containment created by a plantation of trees. **6a** situation of plantation to prevent clean groundwater from entering a contaminated zone; **6b** situation of plantation to prevent the downgradient migration of a contaminated plume

6b) Downgradient of Plume

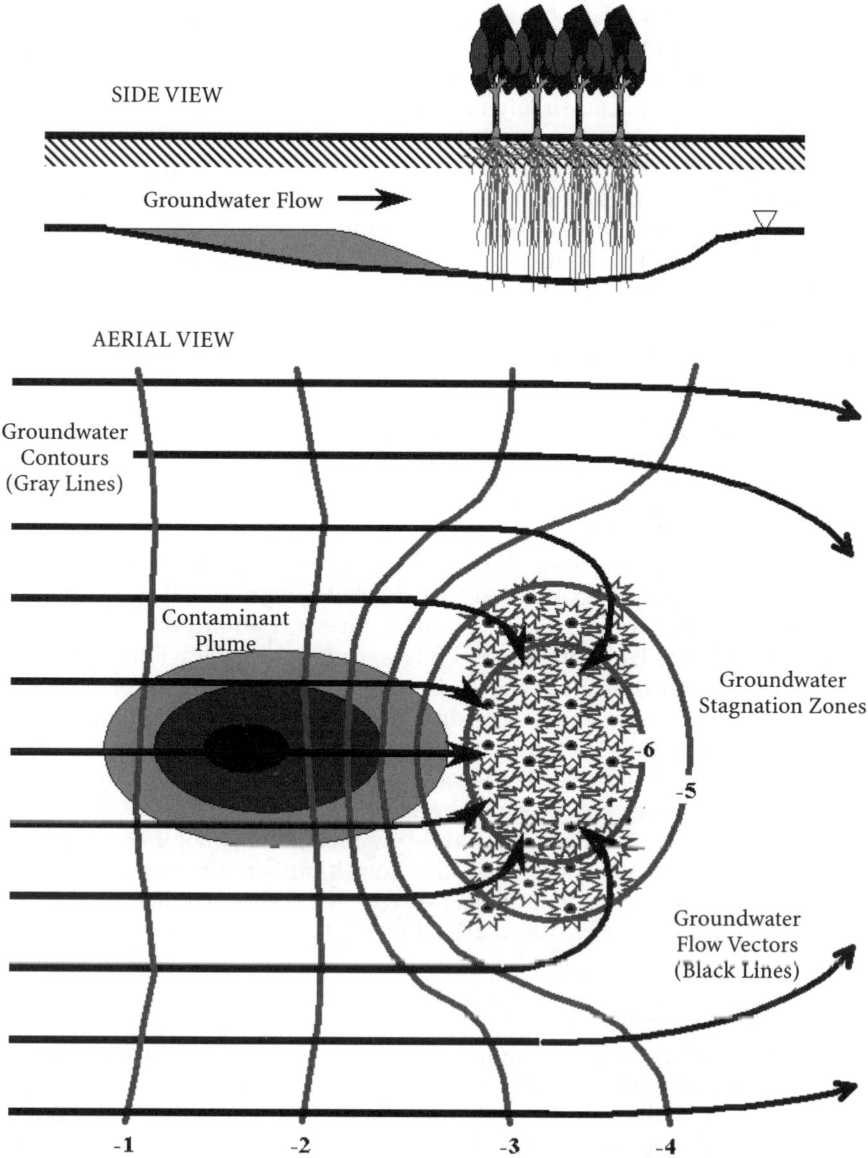

Fig. 6 (continued)

than the direct rainfall amount into the area (f can be greater than one when r is less than zero). Similarly, since rain interception capacities are dependent on the magnitude and duration of the rain event, values for i can vary from 0 to 1. The water that is able to reach the ground surface of the planted area can infiltrate into the shallow root zone and supply the vegetation with a portion of its water demand. As long as V_T for the whole plantation is greater than fR′, and assuming that sufficient plantable area is available, a successful plant hydraulic system can be designed to capture the incoming groundwater (Q). This design criteria is presented in Eq. (8). The supply into the shallow root-zone is dictated not only by the amount of rainfall, but by the soil compaction (which dictates drainage) and evaporative demand. There are, however, engineering control techniques such as soil mounding, impermeable tarps, run-off diversion techniques, and planting designs that encourage deep root systems and can be used to alter plant water usage.

$$R' = A \cdot R \tag{6}$$

$$f = 1 - i - r \tag{7}$$

$$V_T \geq Q + fR' \tag{8}$$

This relatively simplistic water balance for a phytotechnology hydraulic control system does not consider factors such as the daily and seasonal cycles of vegetation, the age of the canopy, root penetration rates, variabilities in climate conditions which dictate ET demands, and fluctuating water tables. Furthermore, the present analysis also assumes that unlimited land is available to plant as much vegetation as necessary to obtain hydraulic control and plume containment.

Furthermore, plume itself must be present at depths that are accessible by the plant roots. Specifically, the overall root depth (D_r) and effective range of influence into a saturated zone (D_u) must be equal to or greater than the plume depth (D_p) and thickness (D_t) in order for this phytotechnology application to be feasible as a stand-alone technology. If this condition is not met, then the deeper portions of the plume may not be captured. This additional design criteria for complete plume capture is summarized in Eq. (9) and shown schematically in Fig. 6.

$$D_p + D_t \leq D_r + D_u \tag{9}$$

Since plant roots require oxygen in order to respire, they generally do not grow into the saturated water table. However, because their rate of water removal is typically greater than the rate of lateral groundwater recharge in the subsurface, they do exert an influence on the local depth of the water table. This zone of groundwater depression is due to the up-welling of water caused by the absorptive capacity of the roots. Therefore, the range of influence of a root system (D_u) is directly related to the rate of transpiration by the specific species. For deep-rooted phreatophytes that typically pump between 20 and 40 gallons per day per 5- to 10-year old tree, the typical range of influence is two feet into the saturated zone [100]. Additional factors that influence D_u include local climate, daily/seasonal cycles, the maturity of the vegetation, and local hydrogeologic conditions (soil permeability, conductivity, grain size, heterogeneity, etc).

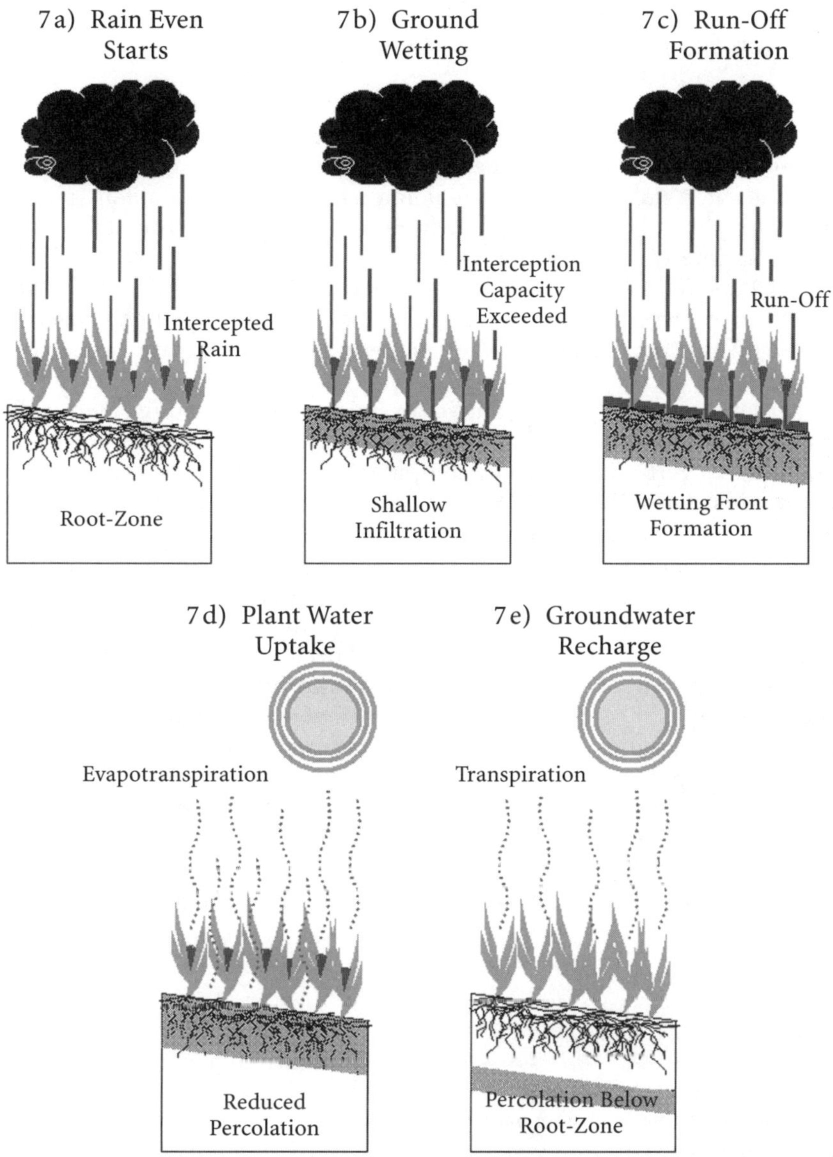

Fig. 7. Steps during a rain event as affected by vegetative covers. **7a** initiation of the rain event where water is intercepted onto the leaves; **7b** the interception capacity is overwhelmed by the volumetric rate; **7c** the formation of run-off as the rain event exceeds the percolation rate; **7d** after the rain event, plant transpiration removes water before percolating below the root-zone; **7e** groundwater recharge

4.1.2
Covers for Infiltration Control

Evapotranspiration covers utilize the capabilities of vegetation to intercept and evaporate rainwater plus transpire soil moisture to prevent water from infiltrating into the subsurface and recharging groundwater. This recharge can lead to a groundwater mound which can potentially cause contaminant plumes to migrate in the subsurface. These covers are specifically designed with optimized rain interception and transpiration capacities of the vegetation (see Section 3.3.2) as well as maximized run-off contouring. Figure 7 illustrates the typical steps occurring during a rain event that is substantial enough to overwhelm the rain interception capacity of the canopy and percolation rate through the soil matrix.

In order to design these vegetative covers, the same equations used to design groundwater hydraulic barriers (Eq. (6) to Eq. (9)) can be used except that Q is reduced to zero. Alternatively, computerized programs such as the Hydrologic Evaluation of Landfill Performance (HELP) model can be used [104]. Since the fraction of rain interception is dependent on the magnitude and duration of the rain event, proper containment of a contaminant plume using these covers requires a good knowledge of the precipitation patterns for the local area. This includes both rain and snow fall and the times and amounts that occur in relation to the growing season. Specifically, design considerations need to include whether there are distinct rain/drought seasons, the volume of water released as snow melt (roughly 8 inches of snow is equivalent to 1 inch of rain), and when the vegetation begins to adequately pump water. Furthermore, the rooting depth, or infiltration capture zone, need to be sufficiently deep in order to capture the water as it percolates downwards. The rate at which water percolates through a soil matrix is dependent on the soil compaction, permeability, and grain size. Each of these soil characteristics also dictate the water holding capacity of the soil which can also be designed to help reduce groundwater recharge.

4.1.3
Erosion Control and Soil/Sediment Stabilization

One of the simplest form of phytotechnologies is the use of vegetation is to provide site stabilization. Soil erosion generated by wind and rain can be prevented using vegetation with root systems that anchor the soil into aggregated clumps. Similarly, sediments can be collected and stabilized behind vegetative barriers to promote soil build up and prevent desertification. Site stabilization applications can also include the phytosequestration mechanism to stabilize inorganics in the rhizosphere. Since these constituents are typically left sequestered, the contaminants are contained and prevented from migrating off site or into groundwater. This application is often referred to as phytostabilization.

Typically, a mixture of grasses and groundcovers are applied to the area requiring stabilization such as slopes and embankments as well as the up- and down-gradient areas. Hydroseeding can be used for larger areas while hand seeding or manual push seeders are reasonable for smaller areas. Several standardized mixes have been developed by local and federal departments of transportation for the stabilization of road side embankments. The composition of these seed mixes usually includes an annual cover crop such as annual fescue (*Festuca* sp.), ryegrass (*Lolium* sp.), and spring oats (*Avena* sp.) to rapidly establish a cover during the initial year of planting plus control volunteer weed growth. Furthermore, these seed mixes also include perennial species such as prairie grasses or common vetches (*Astragalus*, *Coronilla*, *Lotus*, and *Vicia* sp.) that will establish and re-grow annually during subsequent seasons. Seeding rates depend on the individual species in the seed mixtures. Furthermore, a carrier material such as moistened paper fragments or sawdust can be used to aid in the seeding as well as prevent the seeds from being wind blown or scavenged by birds. These materials as well as other forms of mulch such as seedless hay also provide moisture retention that aids germination.

If phytosequestration of an inorganic contaminant is also required, then specialized species capable of sequestering the constituents can be added or substituted for the standard mix. Furthermore, commercially available VAM fungi (ectomycorrhizal) can be added to the soil to aid in the sequestration of inorganics. Regardless of whether inorganics are a concern, the use of VAM fungi is recommended due to the beneficial effects on plant nutrition.

One of the major environmental problems that exists today is the rapid desertification that is occurring worldwide. In many parts of the world, urbanization and deforestation have cause massive erosional problems on a regional scale where arable topsoil is being washed away as surface vegetation is being removed. In many cases, the remaining soils are non-arable due to increase salinity and other inorganics. These problems have lead to reductions in agricultural production and food shortages. In order to combat this problem, soil stabilization can be achieved using deep-rooted grasses planted densely in rows along eroding embankments. The World Bank has advocated the use of non-invasive grasses such as vetiver (*Vetiveria* sp.) and related species (*Cymbopogon*, *Panicum*, *Stipa*, other sp.) [105] to accomplish these tasks. The basic set up for these systems is to plant rows of these grasses (which eventually forms thick hedges up to 2 m tall) along the equilateral contours of a sloping area. These grow so dense that they form (very rapidly given the correct climate) "botanical dams." These dams prevent the loss of topsoil and sediments, thus, reducing erosion and desertification. They also promote the seepage of water into the subsurface as opposed to forming run-off. Unlike earthen dams, these are capable of allowing water to flow through while retaining the sediments; therefore, ponding/flooding is also less of a problem with these systems. In places where desertification has already occurred, the use of these long-lived, disease resistant species can result in the build-up of 50–60 cm of sediments within the first year [105]. Furthermore, several of these systems have been in operation for over 50 years and have survived several adverse environmental conditions.

4.2 Treatment Strategies

Strategies designed for treating or remediating contaminants from impacted sites generally include one or a combination of rhizodegradation, phytodegradation, phytoextraction, or phytovolatilization. Since phytovolatilization may or may not be considered acceptable by regulators, most applications do not utilize this as a primary clean up strategy. Typically, this is monitored as a potential side effect of the application, performed in conjunction with site risk or air quality assessments. Treatment strategies can be applied to soils, sediments, and surface water. Groundwater can also be treated in these applications; however, these treatment strategies are typically combined with containment objectives. The applications with dual objectives of containment and treatment are covered in the following section (see Sect. 4.3).

4.2.1 *Surface Soil Clean Up*

Surface soils impacted with organic, inorganic, or a mixed contaminant situation can be seeded with appropriate vegetation in order to cleanse the soil. For organics, the primary mechanism is rhizodegradation although phytodegradation can also be a contributing clean up mechanism. In these applications, the typical design depths that can be treated are less than two feet. The species selected for these applications contain root systems that are fibrous and dense in order to maximum the volume of the rhizosphere. Candidate species for these applications include grasses such as fescue (*Festuca* sp.), ryegrass (*Lolium* sp.), and wheatgrass (*Agropyron* sp.) as well as some herbaceous species including clovers (*Trifolium* sp.) and other legumes [27, 106]. However, deeper rooted species such as many of those illustrated in Fig. 4 can be used to treat organically impacted soils at depth. Furthermore, woody vegetation which contain deeper root systems can be applied as well. During the natural regrowth of an old sludge basin, root penetrations down to five feet were achieved [25].

In some cases, these phytotechnology applications can be combined with other techniques designed to stimulate microbial biodegradation. High concentration areas (known as hot spots) can be removed using conventional excavation and removal while the larger, less contaminated areas can be vegetated to promote rhizodegradation. In a landfarm situation, initial tilling is effective at reducing the volatile fractions of organic contaminants. When this has reached its (asymptotic) limits of effectiveness or when constituents leach down below the zone of tilling, phytotechnologies can be applied to address the semi-volatile and leached components [51].

When addressing inorganic contaminants in surface soils, hyperaccumulators or halophytes are generally seeded onto the impacted area in order to phytoextract the constituents. Certain inorganic contaminants are readily or moderately available to be taken up into plant tissues including As, Cd, Cu, Fe, Mn, Ni, Se, and Zn. Other inorganics such as Cr, Pb, and U are less available due to low solubilities and generally do not enter into the plant tissues at appreciable quantities. In these cases, soil amendments can be applied to aid in the uptake of the inorganics into the terrestrial portion of the plant tissues. These amendments can include chelating or solubilizing agents such as ethylene diamine tetraacetic acid (EDTA), citric acid, and ammonium nitrate [2, 107–109]. Since these amendments can lead to excessive leaching of the contaminants deeper into the subsurface, their application is designed to correspond to the optimum time when the vegetation is rapidly taking up soil moisture. Furthermore, in order to discourage the transfer into the food chain, these applications are also conducted before the vegetation enters into the seed production phase of growth.

Once the vegetation has reached a certain growth stage, the terrestrial tissues are typically harvested and dry ashed or burned to remove the bulk of the organic biomass. Leaving behind an inorganic-rich ash, this residue can be sampled and then disposed of appropriately in a waste management facility. The estimated reduction in overall mass disposed can result in savings of over 70% [110]. Although this represents a major cost-savings potential, the need for multiple harvests can restrict the economy of this phytotechnology application. Therefore, in addition to selecting species that are able to hyperaccumulate the constituents of interest, high and rapid biomass production are also desirable criteria. Current research developments include the potential genetic enhancement of known hyperaccumulators to produce increased biomass.

Since the vegetation is subjected to adverse environmental conditions and the survival of the vegetation, to a certain extent, is paramount to the success of the phytotechnology application, then maintaining some control of the growth conditions at the impacted site is necessary. In general, vegetation will respond favorably to periodic fertilization and irrigation. Fertilization can be specially defined for the site using agronomic data including inorganic nutrition (see Section 2.2.1), organic matter content, soil moisture, acidity, alkalinity, electrical conductivity, and cation exchange capacity. General 10–10–10 (N-P-K) fertilizers can be used at roughly 10 lbs per 1000 ft^2 for sites that are less defined. Similarly, irrigation requirements can be between 1 and 1.5 inches per week in applications conducted 2 to 3 times per week. The best time to apply irrigation is at dusk so that the water can soak in deeper into the soils before evapotranspiration reaches its maximum during the following day. This thorough soaking also promotes deeper root systems to some extent. In many phytotechnology applications, automated irrigation systems can be set up.

4.2.2
Treatment Wetlands for Sediment/Surface Water Clean Up

Treatment wetland systems is a broad category of aquatic systems that utilize vegetation in combination with flow control systems to treat and contain environmental contaminants present in surface waters and sediments. Besides the typical treatment wetland, this application of phytotechnologies also includes vegetated swales, ponds, lagoons, and reedbeds. This area of phytotechnology is perhaps the oldest in terms of applications in the field and technical development. Some systems have been in operation for several decades to treat industrial as well as municipal wastewaters, stormwater, and run-off [76].

In addition to the various phytotechnology mechanisms, constructed wetland treatment systems contain several physical and chemical mechanisms that can contribute to the treatment of environmental contaminants. These mechanisms include volatilization, settling and sedimentation, adsorption, chemical precipitation, and photo-oxidation. Furthermore, the biological activity occurring in the sediments themselves promotes aerobic and anaerobic biodegradation. In the sediment layer of a wetland system, a relatively thin aerobic zone overlays the anaerobic zone located deeper in the sediments. The depth of the aerobic and anaerobic layers depends on the rate of sediment build-up, turbulence or quiescence of the water flowing above, presence of emergent or submerged vegetation, and interactions from amphibians, avians, mammals, benthic organisms, or other aquatic animals.

In addition to typical environmental contaminants such as metals and organic chemicals, constructed treatment wetlands can also be used to mitigate pathogenic organisms such as *E. coli*, *Streptococci*, *Salmonella*, and fecal coliforms. Similarly, many industrial wastewater treatment plants (WWTP) and publicly operated treatment works (POTW) use these systems to meet other water quality standards including turbidity, total suspended solids (TSS), total dissolved solids (TDS), chemical oxygen demand (COD), five-day biochemical oxygen demand (BOD_5), and total organic carbon (TOC). Most of these parameters need to meet some sort of discharge permit requirement such as the National Pollution Discharge Elimination System (NPDES) permits under the US Clean Water Act (Section 402). These permits can be based on concentration levels (mass per unit volume) and/or total mass loads over a given time frame (mass per unit time). Reported removal efficiencies for various water quality parameters are listed in Tables 2–5 [76, 111, 112].

The transpiration capacity of wetland species (see Fig. 5) as well as the evaporative losses from free surface water can be utilized to control and contain surface water flows. Furthermore, the design of the treatment wetland system can also be optimized to control hydraulics by including cascading systems, air spargers, retention/detention basins, baffles, float controls, and liners. Furthermore, different types of flow systems can be designed including surface flow (SF) systems, subsurface flow (SSF) systems, and vertical flow (VF) systems. The hydrodynamics of each of these designs are shown in Fig. 8.

The advantages of SF wetland systems include the relative ease to construct, simple designs, less costly installation, a high degree of flow control, and habitat value. The disadvantages include odor generation, pests, potential exposures, and a somewhat restricted list of species that can be utilized (typically OBL and FACW species only). Wetland species only survive in certain depths of water and depend on whether they are submerged or emergent species [76]. These depths can range to over 1.5 m but typically are less than 0.5 m. Some species have developed physical characteristics such as air roots, lenticels, and aerenchymous tissues that allow them to withstand deeper waters. Since plant depths limit this dimension of SF systems (z), then designs based on a dynamic water balance (change in volume over time, dV/dt) must be of an appropriate surficial area (A). However, portions of a wetland system can contain non-vegetated ponds or basins for increased holding capacity. The components of a water balance are

Table 2. Reported removal efficiencies for other water quality parameters in constructed treatment wetland systems

Water quality parameter	Representative group/industry	Inlet concentration range		Units	% Concentration reduction range		Inlet mass (kg/ha-day)	% Mass reduction
BOD_5	North American		29.8	mg/L		73	10.9	68
	Pulp & Paper		26.1			48	28.6	29
	Livestock	104	442		25	83		
	Petroleum	1.9	700		5	98		
COD	Petroleum	101	3150	mg/L	30	96		
TSS	North American		46.0	mg/L		72	16.8	71
	Pulp & Paper		42.5			71	41.6	68
	Livestock	128	1111		47	81		
	Petroleum	1.3	181		–362	89		
Fecal coliforms		23.9	7.7×10^5	col/L	–182	99.9		
Total coliforms		8.98	6750	col/1000 L	94.8	99.1		
Fecal streptococcus		80	9000	col/L	66.1	99.8		
Color	Pulp & Paper	1617	1617	mg/L		2	2541	–4

North American Group based on average values from the North American Treatment Wetland Database.

Table 3. Reported removal efficiencies for inorganic macroelements in constructed treatment wetland systems

Inorganic macroelements	Representative group/industry	Inlet concentration range (mg/L)		% Concentration reduction range		Inlet mass rate range (kg/ha-day)		% Mass reduction range	
NH_4	North American		4.97		52		1.46		26
	Pulp & Paper		4.7		36		3.6		14
	Livestock	5.1	366	20	60				
	Petroleum	0.2	16.9	40	98				
NO_2+NO_3	North American		5.49		62		0.99		55
	Pulp & Paper		1.4		90		0.49		86
NO_3	Petroleum	1.1	5.9	68	97				
Organic N	North American		4.01		49		1.71		56
TKN	North American		8.11		44		2.99		43
	Pulp & Paper		7.8		55		3.6		31
N (total)	North American		9.67		53		2.98		51
	Pulp & Paper		12.6		48		4.2		38
	Livestock	89	407	22	51				
	Petroleum	3.2	16.2	21	99				
P (organic)	North American		1.75		37		0.29		41
P (total)	North American		3.80		56		0.73		31
	Pulp & Paper		2.3		26		1.0		
	Petroleum	0.08	98.7	−94	81				
K						0.003	10.13	−255	45
Fe		0.1	205	−217	98	0.009	84.5	−135	98
Na						0.012	3.32	− 78	43

North American Group based on average values from the North American Treatment Wetland Database.

Table 4. Reported removal efficiencies for inorganic microelements in constructed treatment wetland systems

Inorganic microelements	Inlet concentration range (μg/l)		% Concentration reduction range		Inlet mass rate range (kg/ha-yr)		% Mass reduction range	
Al	0.3	300	–33	67	1.1	45.3	–66	63
Cd	0.1	70	0	99.7	0.08	75.1	5.9	99
Cr	< 2	160	0	88	0.23	22.3	68	96
Cu	1.4	1510	36	96	0.81	268	40	96
Pb	1.7	300	–181	98	0.03	66.9	–107	86
Mn	210	7400	40	98	0.26	1113	7.7	79
Hg	0.02	5.7	17	66	0.002	51	4.8	18
Ni	4.0	300	25	90	0.22	66.9	49	71
Zn	0.01	6900	33	98	0.27	558	33	91

Table 5. Reported removal efficiencies for organic constituents in constructed treatment wetland systems

Organic constituents	Inlet concentration range		Units	% Concentration reduction range	
O&G	0.84	24	mg/L	54	94
Phenols	6.2	80	μg/L	10	94
VOC		32.2	μg/L	98	98
SVOC		5.0	μg/L	84	84

similar to those for hydraulic barriers (see Eqs. (6) to (8)) including wastewater input (Q), precipitation ($R'=AR$), run-off (typically run-on; $r<1$), infiltration (I), and evapotranspiration (ET) [76]. Note that since rain interception (i in Eq. (7)) leads to evaporation, this is combined with the evaporation occurring from the free water surfaces in the SF system to generate the overall evaporation. In most cases, the total water losses, which combines total evaporation and plant transpiration into evapotranspiration, are estimated empirically from lined wetland systems ($I=0$).

$$dV/dt = A\,(dz/dt) = [Q + R'(1-r)] - [I + ET] \tag{10}$$

Alternatively, the advantages of SSF systems include odor control, increased accessibility, reduction in pests, decreased exposure risks, a more robust operation in the winter months due to insulation, and a wider range of vegetation that can be utilized (OBL, FACW, FAC, and FACU species). However, SSF systems are prone to clogging, require more engineering and design, and are harder and more expensive to install. Typical SSF systems utilize high permeability media such as pea gravel, cobble, rip-rap, or similar material that allow the continuous flow of water through the system. These are generally overlaid with an appropriate medium that allows root penetration into the high permeability material as well. Since the volume of these systems is typically restricted, diversion systems (baffles, weirs, etc.) can be used to capture the "first flush" of a large rain event and then allow

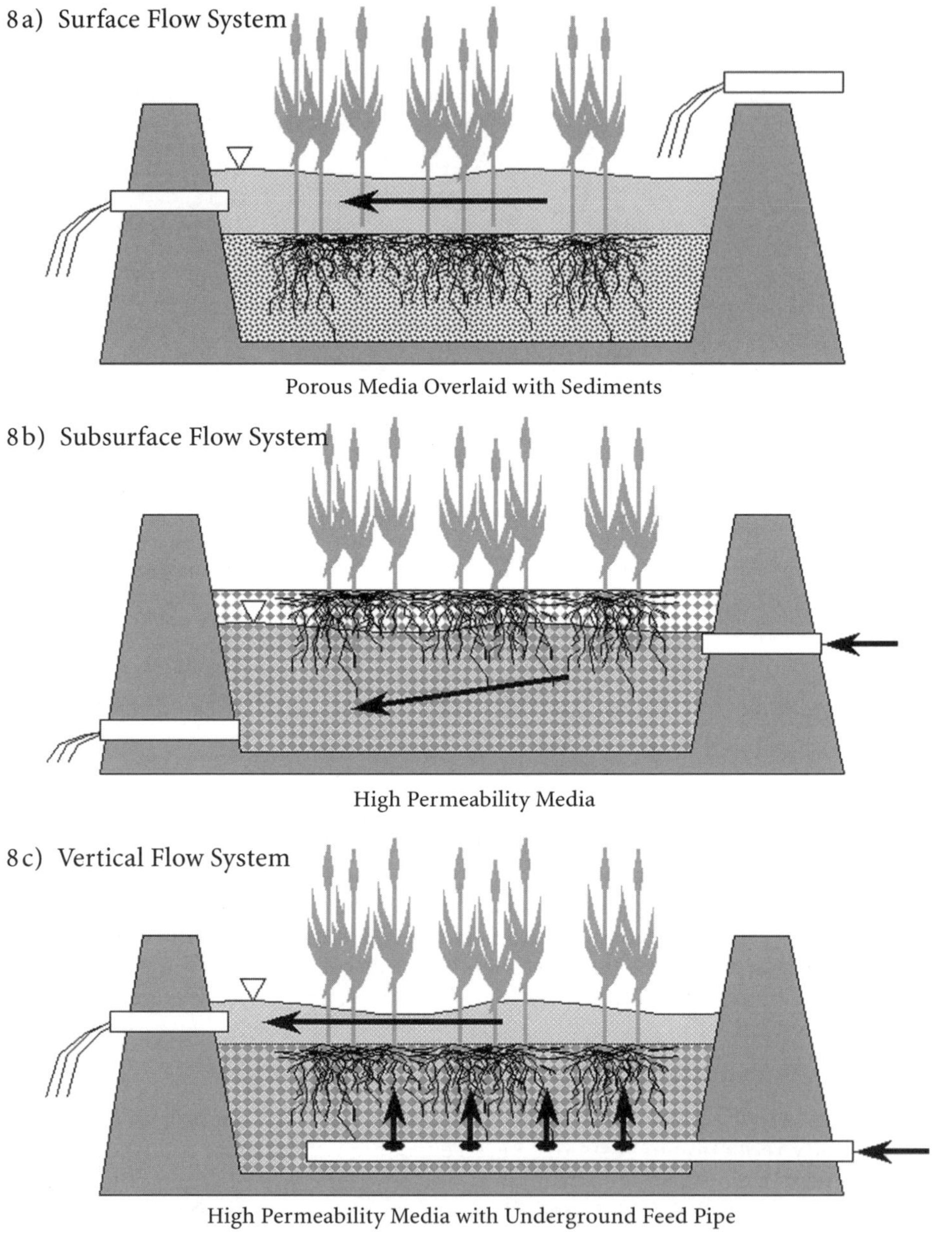

Fig. 8. The basic hydrodynamic characteristics of constructed treatment wetlands. **8a** surface flow systems, **8b** subsurface flow systems, **8c** vertical flow systems

the remainder of the rainwater to be captured in an area designed for extra storage capacity [111]. It is this first flush that generally contains much of the dissolved and suspended constituents that will be generated during the rain event.

A variant of these systems involves the use of underground piping that can be used to promote the vertical flow of water through the system. The primary advantage of including this additional plumbing is the ability to increase the contact time of the constituents to be treated through the highly active rhizosphere of the wetland vegetation. Furthermore, oxygen can also be introduced into the root-zone of the vegetation in order to maintain a larger aerobic zone in the sediments. The primary disadvantage of these VF systems is the additional expense to install the plumbing.

In some instances, it may be reasonable to combine SF, SSF, and VF systems into a single treatment system by dividing the area into cells. Furthermore, these can be combined with pond or lagoon cells to increase the water holding capacity so that a higher degree of flow control can be obtained or more than the first flush can be captured if required. Each of these cells can be maintained in continuous operation or can be isolated using the flow diversion techniques to be used as required. The water balance in Eq. (10) can be determined for each individual cell as well as the entire wetland treatment system. The disadvantage of these highly dynamic systems is the increased land area required.

The three main types of vegetation used in many constructed wetland systems include cattails (*Typha* sp.), bulrushes (*Scirpus* sp.), and reeds (*Phragmites* sp.). Each of these are OBL to FACW species. In some areas these species may be considered noxious or invasive and may not be appropriate for the site. Other species which have been used in these treatment systems, which range from OBL to FAC on the wetland indicator status, include waterweed (*Elodea canadensis*), duckweed (*Lemna minor*), parrot feather (*Myriophyllum aquaticum*), reed canarygrass (*Phalaris arundinaceae*), bamboos (*Bambusa* sp.), sedges (*Carex* sp.), and pondweeds (*Potamogeton* sp.) [34, 70, 76, 113–117]. Again, species selection will depends on many factors common to all phytotechnology applications including climate, ease of planting, economics, availability, remedial effectiveness, and habitat value. In addition to these common criteria, the species selected for constructed treatment wetlands also need to consider the operation in the winter (ice formation), disruptions caused by large storm events or extended droughts, and the potential generation of additional water quality conditions due to biological use (i. e. fecal deposition by water fowl).

4.3 Containment and Treatment Strategies

Several phytotechnology applications combine the capabilities of vegetation to provide hydraulic control with the abilities to treat contaminant materials. In fact, in many applications, containment and treatment are difficult to separate since water uptake brings dissolved constituents into the highly bioactive rhizosphere. The aqueous media that these combined strategies can be applied include both surface water and groundwater.

4.3.1
Riparian Buffers for Surface Water Protection

Surface water bodies can be impacted from non-point sources of pollution derived from run-off from road ways, agricultural fields, and urban areas. In addition to dissolved chemicals such as road salts, agrochemicals, animal wastes, and oils from leaking vehicles, particulate matter and sediments can be carried into the water body as well. Similarly, in many geological settings, surface water bodies are hydraulically connected to the local groundwater system or aquifer. Therefore, groundwater impacts can eventually find their way into the surface water body as well. In order to protection these waters, riparian buffers can be established along the boundaries. These buffers are vegetated strips or transitions designed so that dissolved constituents can infiltrate into the rhizosphere and suspended solids can settle out prior to reaching the water body. Also known as corridors, these systems utilize the various phytotechnology mechanisms to cleanse and/or contain the run-off.

These systems are typically situated parallel to the banks of the water body to be protected along the entire length. In the case of rivers or streams, these systems can be very long. The design of riparian buffers is based on the hydraulic

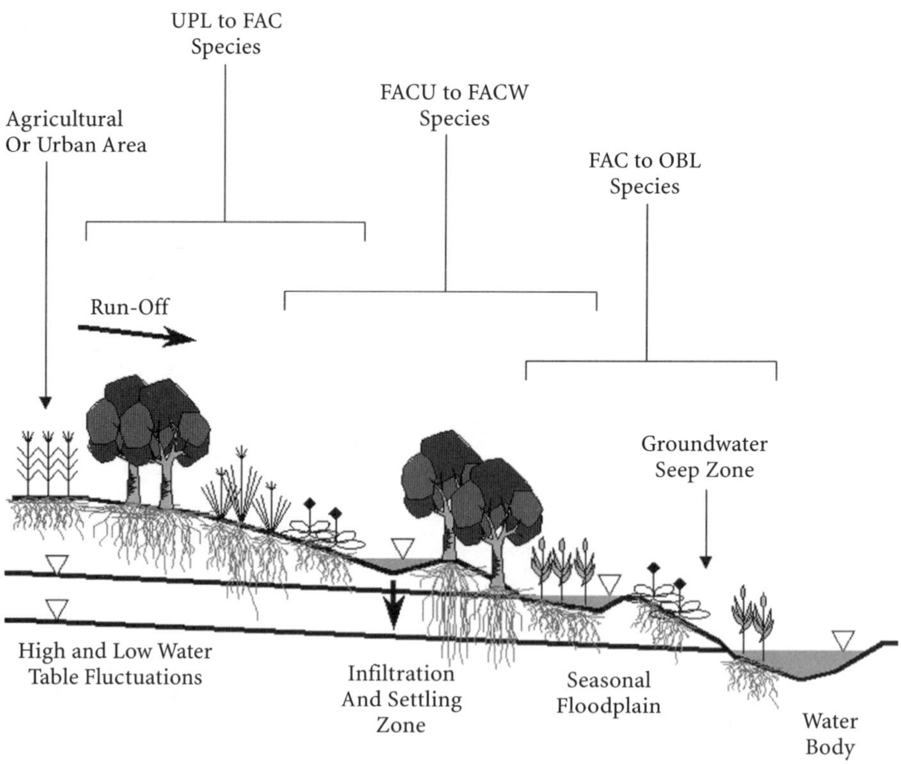

Fig. 9. Cross-sectional schematic of a riparian buffer system

load coming into the vegetated area. For suspended matter, sufficient residence time must be provided in order for settling and sedimentation to occur. This can be facilitated by creating small berms or swales along the length of the system. These parallel rows allow some ponding and infiltration to occur so that the non-point source pollution is captured and then introduced into the highly bioactive rhizosphere of the vegetation. Once in the rhizosphere, phytosequestration, rhizodegradation, phytoextraction, phytodegradation, phytovolatilization, and evapotranspiration can lead to the containment or remediation of the constituents. A cross-sectional view of a riparian buffer is shown in Fig. 9.

Similar to wetland treatment systems, riparian buffers can utilize all classes of vegetation on the wetland indicator status. Closest to the water body, which can typically be within the flood plain, OBL and FACW species can be used. Moving upland, vegetation in the transition between saturated and unsaturated soils can include FACW, FAC, and FACU species. Finally, in the upland areas closest to the non-point source pollution, FACU and UPL species can be planted.

Table 6. Common species used in phytotechnology applications based on the wetland indicator status

Wetland species (OBL, FACW):	*Betula nigra* (river birch) *Carex* sp. (sedges) *Myriophyllum aquaticum* (parrot feather) *Phalaris arundinaceae* (reed canarygrass) *Phragmites* sp. (reeds) *Scirpus* sp. (bulrushes) *Spartina* sp. (cordgrasses) *Taxodium distichum* (bald cypress) *Typha* sp. (cattails)
Facultative species (FACW, FAC, FACU):	*Andropogon* sp. (bluestem grasses) *Celtis occidentalis* (hackberry) *Elaeagnus augustifolia* (russian olive) *Eucalyptus camaldulensis* (river redgum) *Festuca rubra* (red fescue) *Lolium* sp. (ryegrasses) *Lotus corniculatus* (birdsfoot trefoil) *Morus rubra* (mulberry) *Populus* sp. (poplars) *Populus deltoides* (cottonwoods) *Salix* sp. (willows) *Trifolium hybridum* (alsike clover) *Trifolium repens* (white clover)
Upland species (FACU, UPL):	*Agropyron* sp. (wheatgrasses) *Bouteloua* sp. (grama grasses) *Coronilla varia* (crownvetch) *Festuca arundinacea* (tall fescue) *Helianthus annuus* (common sunflower) *Panicum virgatum* (switchgrass) *Pinus taeda* (loblolly pines) *Solidago rigida* (stiff goldenrod) *Trifolium pratense* (red clover)

A list of common phytotechnology species, based on their wetland indicator status, is provided in Table 6. When treating groundwater seeps, deeper rooted species may need to be selected, particularly on the upland side of the riparian buffer.

4.3.2
In-Situ Groundwater/Subsurface Soil Treatment

Relatively shallow aquifers and the corresponding subsurface soils can be treated using dense stands of deep-rooted vegetation. Again, the various phytotechnology mechanisms are utilized to contain (primarily through evapotranspiration) and treat the contaminants. Typically, phreatophytic trees are used in these applications although other deep-rooted species such as prairie grasses can be used as well. Aquifers within the rooting depths normal for phreatophytes (10 to 15 ft bgs [36]) can be planted using several available methods. The easiest technique in sandy or loose soils is to simply poke small, individual holes using a rod or dowel into the soil, place a plug or whip (1 to 2 foot tall stock) into the hole, and follow by hand packing the soil back around the stock. The advantage of this method is that large plantations can be established in a relatively short time (experienced forestry personnel can plant several thousand trees per day). Alternatively, for clayey or tighter soils, whips can also be planted using a trench digger followed by personnel to manually backfill the trench. In this case, several hundred whips can be planted by experienced personnel per day. The disadvantage of these two techniques is that the trees will require substantial time in order to produce roots that will tap into the groundwater at depth.

For larger stock such as poles (up to ten or so feet in length), individual boreholes can be augered in using a drill rig or similar auger mechanism. Typically, a six to twelve inch auger is used to produce a borehole down to groundwater. The advantage of this technique is that the larger stock can be "deep planted" which utilizes the characteristics of certain phreatophytes that contain pre-formed growth initials (primordia) located on nodes along the length of the pole [118]. Those initials that are planted below ground produce roots while those located above ground produce branches and leaves. Basically, this characteristic of the trees allows the poles to be planted closer to the saturated zone which decreases the time needed to begin tapping into the groundwater.

After the boreholes are excavated, backfilling with a sand:peat mixture (roughly 60 : 40 mix) can be conducted to create a conduit that will provide preferential root growth pathways, particularly in clayey soils. Assuming the excavated soils cannot be classified as a waste material (based on local regulations), the backfill mixture can also include a portion of the native material to cut down the volume of sand and peat required and subsequent disposal of the excess soil. In order to insure adequate root growth and develop, general fertilizers can be added to the backfill mixture in powder form or as commercially available slow-release fertilizer tablets. Using this technique, up to one hundred trees can be planted per day; although, this depends on the depths needed to be excavated.

Since plant roots require oxygen and moisture, breather tubes (typically one inch, perforated polyvinyl chloride tubes) and irrigation systems can also be installed. The breather tubes allow atmospheric oxygen to diffuse into the subsurface and are placed into the borehole prior to backfilling. Furthermore, deeper root growth can be promoted using sub-irrigation systems which basically consist of individual drop-down tubes (i.e. Tygon tubing) from which irrigation is supplied into the subsurface. These tubes are also installed into the borehole prior to backfilling. In some cases, multi-level sub-irrigation systems can be used to progressively increase the depths at which irrigation is supplied. This causes the roots to penetrate progressively deeper in order to obtain moisture before tapping in the groundwater. The disadvantages of utilizing these cultivation systems are the added costs involved. Additional techniques to aid in deeper root penetration include soil mounding, installing plastic surface material, and planting grasses or groundcover vegetation around the base of the trees. These techniques reduce the amount of surface infiltration, causing the trees to seek deeper for their moisture requirements.

There is also a patented technique which utilizes plastic collars situated along the entire length of the borehole down to groundwater [119]. The basic concept behind this special technique is that the collars prevent lateral root growth from the tree planted within the collar while forcing the roots to grow vertically down the borehole. Using this technique, treatment of groundwater as deep as 35 feet has been attempted [119].

The typical spacing between trees in a row can range from 5 to 10 feet with neighboring rows offset so that adjacent trees are at an angle of 60 degrees (see Fig. 6). At five foot spacings, this translates into 20 ft^2 of ground area per tree while at ten foot spacing, this equals approximately 75 ft^2 per tree. At some sites, trees are being planted as close as one foot apart (approximately 1 ft^2 per tree) with the intention that thinning will occur both naturally and during the normally operation and maintenance of the site. Using these area per tree values, the total number of trees that can be planted on the available areas of the site can be determined. Furthermore, initial designs should use conservative water usage rates (see Fig. 5 and Sect. 4.1.1) in order to calculate the overall uptake occurring from the stand. One consideration that may need to be included is that the growth of the vegetation may be stunted due to the adverse site conditions which could correspond to a reduction in transpiration rates. Furthermore, 10–20% mortality may be expected simply due to the transplantation of the vegetation rather than any toxic effects from the contamination. Therefore, when developing the initial budget for these phytotechnology applications, a re-planting contingency should be incorporated.

4.3.3
Ex-Situ Groundwater Rhizofiltration

At many environmentally impacted sites, groundwater extraction systems are already in operation. Alternatively, due to the depth of the impacted groundwater, physical extraction systems are the only means of accessing the contaminant plumes. After this water is brought up to the surface, treatment systems are used to cleanse the water to discharge conditions. These are known as pump and treat (P & T) systems.

In some cases, above-ground plant hydroponic systems can be used as the treatment portion of the P & T system where the extracted groundwater is passed through the root systems of the hydroponically grown plants. This application has been defined in phytotechnologies as rhizofiltration [120]. Specific applications have been utilized for filtering inorganics through the root systems of these hydroponically cultivated plants. The primary phytotechnology mechanism utilized is phytosequestration (on or in the root tissues) although phytoextraction can occur as well. Although no current applications of rhizofiltration have been developed for organic contaminants, these systems can be designed to treat these constituents through phytodegradation. Since there is no true soil medium in these hydroponic systems, rhizodegradation cannot be used as a primary clean up mechanism.

Typical hydroponic systems grow plants without a soil medium although an artificial medium can certainly be utilized. These can include sand, gravel, perlite, vermiculite, plastic beads, ceramic materials, or other permeable media [17]. In this sense, this application is very similar to a subsurface treatment wetland system (see Sect. 4.2.2). Those systems designed to treat pumped contaminated groundwater without a rooting medium allow for the easy harvest of the plant material that contain high concentrations of the contaminant material. Plant support can be established using wire mesh screens or perforated styrofoam blocks that float on the aqueous stream. Because of size restrictions in developing economical large-scale hydroponic systems, rhizofiltration systems are generally applied to low-flow P & T systems. In one case, floating styrofoam rhizofiltration systems were applied onto an impacted water body allowing passive phytosequestration to occur [120, 121]. The design of these systems require a rate of sequestration to be developed in order to balance the rate of volumetric flow through the system. Once the volumetric rate is defined, the size (volume) of the hydroponic system can be calculated. In some applications, rhizofiltration systems can be equipped with pH adjustment units to promote precipitation as well as small air sparging stones to promote oxidation and precipitation.

5 References

1. Shimp JF, Tracy JC, Davis LC, Lee E, Huang W, Erickson LE (1993) Environ Sci Technol 23:41
2. Schnoor, JL (1998) Phytoremediation. Technology Evaluation Report TE-98-01 Ground-Water Remediation Technologies Analysis Center (GWRTAC)
3. Dragun J (1998) The Soil Chemistry of Hazardous Materials, 2nd ed, Amherst Scientific Publishers, MA
4. Anderson TA, Guthrie EA, Walton BT (1993) Environ Sci Technol 27:2630
5. Reynolds CM, Pidgeon CS, Perry LB, Koenen BA, Pelton DK, Nichols HL, Wolf DC (1999) Using Microbial Community Structure Changes to Evaluate Phytoremediation. In: Leeson A, Alleman BC (eds) Phytoremediation and Innovative Strategies for Specialized Remedial Applications, Vol. 5(6), Proceedings of the Fifth International In Situ and On-Site Bioremediation Symposium, Battelle Press, OH, p 33

6. Azadpour-Keeley A, Russell HH, Sewell, GW (1999) Microbial Processes Affecting Monitored Natural Attenuation of Contaminants in the Subsurface. Ground Water Issue EPA/540/S-99/001
7. White DC, Ringelberg DB, Hedrick DB, Nivens DE (1993) Rapid Identification of Microbes from Clinical and Environmental Matrices by Mass Spectroscopy. In: Fenselau C (ed) Identification of Microorganisms by Mass Spectroscopy, ACS Symposium Series 541, American Chemical Society, Washington, DC, p 8
8. MacNaughton SJ, Stephen JR, Venosa AD, Davis GA, Chang YJ, White DC (1999) Appl Environ Microbiol 65(8):3566
9. Howard PH, Boethling RS, Jarvis WF, Meylan WM, Michalenko EM (1991) Handbook of Environmental Degradation Rates. Lewis , MI
10. McAllister PM, Chiang CY (1994) Ground Water Monit Rem 14(2):161
11. Hanson RS, Hanson TE (1996) Microbiol Rev 60(2):439
12. Steffan RJ, McClay K, Vainberg S, Condee CW, Zhang D (1997) Appl Environ Microbiol 63(11):4216
13. Sylvia DM, Williams SE (1992) Vesicular-Arbuscular Mycorrhizae and Environmental Stress. In: Bethlenfalvay GJ, Linderman RG (eds) Mycorrhizae in Sustainable Agriculture, ASA Special Publication Number 54, p 101
14. Bethlenfalvay GJ, Linderman RG (1992) Preface. In: Bethlenfalvay GJ, Linderman RG (eds) Mycorrhizae in Sustainable Agriculture, ASA Special Publication Number 54, p viii
15. Reid CPP (1990) Mycorrhizae: A Soil-Root Interface in Plant Nutrition. In: Todd RL, Giddens JE (eds) Microbial-Plant Interactions, ASA Publication Number 47, p 29
16. Taiz L, Zeiger E (1991) Plant Physiology, Benjamin/Cummings, California
17. Tsao DT (1997) Modeling the Environmentally and Biologically Controlled Nutrient Uptake Kinetics into Plants. PhD Thesis, Purdue University, IN
18. US EPA (1999) Phytoremediation Resource Guide. EPA Document # 542-B-99-003, U.S. Environmental Protection Agency Solid Waste and Emergency Response (5102G)
19. Koch I, Wang L, Ollson CA, Cullen WR, Reimer KJ (2000) Environ Sci Technol 34:22
20. Trejo CL, Clephan AL, Davies, WJ (1995) Plant Physiol 109:803
21. Elliott LF, Gilmour CM, Lynch JM, Tittemore D (1990) Bacterial Colonization of Plant Roots. In: Todd RL, Giddens JE (eds) Microbial-Plant Interactions, ASA Publication Number 47, p 1
22. Hedge RS, Fletcher JS (1996) Chemosphere 32:2471
23. Tsao DT (1990) The Elicited and Enhanced Production of Phytoalexins in Cotton Suspension Cultures. MS Thesis, Purdue University, IN
24. Flint HL (1997) Landscape Plants for Eastern North America, 2nd ed, John Wiley, NY
25. Olsen PE, Fletcher JS (1999) Bioremediation J 3(1):27
26. Leigh MB, Fletcher JS, Kyle MD, Nagle DP, Fu X, Schmitz FJ (2000) Phytoremediation State of the Science Conference, Boston, MA
27. Ferro AM, Sims RC, Bugbee B (1994) J Environ Qual 23:272
28. Fletcher JS, Donnelly PK, Hedge RS (1995) Biostimulation of PCB-Degrading Bacteria by Compounds Released from Plant Roots. In: Bioremediation of Recalcitrant Organics, Battelle, Columbus, OH, p 131
29. Aprill W, Sims RC (1990) Chemosphere 20:253
30. Schwab AP, Banks MK (1994) Biologically Mediated Dissipation of Polyaromatic Hydrocarbons in the Root Zone. In: Anderson TA, Coats JR (eds) Bioremediation through Rhizosphere Technology. ACS Symposium Series 563, p 132
31. US AEC (1998) US Army Environmental Center Pollution Prevention Environmental Technology Division Report No. SFIM-AEC-ET-CR-97059
32. McCutcheon S (1996) Phytoremediation of Organic Compounds: Science Validation and Field Testing. In: Kovalick WW, Olexsey R (eds) US EPA Workshop on Phytoremediation of Organic Wastes, Ft. Worth, TX
33. Newman LE, Strand SE, Choe N, Duffy J, Ekuan G, Ruszaj M, Shurtleff BB, Wilmoth J, Heilman P, Gordon MP (1997) Environ Sci Technol 31:1062

34. Nzengung VA, Wolfe LN, Rennels DE, McCutcheon SC, Wang C (1999) Internat J Phytorem 1(3):203
35. Flanders C, Dec J, Bollag J-M (1999) Bioremediation J 3(4):315
36. US EPA (2000) Introduction to Phytoremediation. EPA Document # EPA/600/R-99/107. U.S. Environmental Protection Agency Office of Research and Development, Washington, DC
37. Howard PH, Jarvis WF, Sage GW, Basu DK, Gray DA, Meylan WM, Crosbie EK (1990a) Handbook of Environmental Fate and Exposure Data for Organic Chemicals, Volume I: Large Production and Priority Pollutants, 2nd edn. Lewis, Chelsea, MI
38. Howard PH, Sage GW, Jarvis WF, Gray DA (1990b) Handbook of Environmental Fate and Exposure Data for Organic Chemicals, Volume II: Solvents, 2nd edn. Lewis, Chelsea, MI
39. Howard PH, Michalenko EM, Jarvis WF, Basu DK, Sage GW, Meylan WM, Beauman JA, Gray DA (1990c) Handbook of Environmental Fate and Exposure Data for Organic Chemicals, Volume III: Pesticides, 2nd edn. Lewis, Chelsea, MI
40. Turpeinen R, Salminen J, Kairesalo T (2000) Environ Sci Technol 34:5152
41. Salt DE, Blaylock M, Nanda Kumar PBA, Dushenkov V, Ensley BD, Chet I, Raskin I (1995) Biotechnol 13:468
42. Volesky B, Holan ZR (1995) Biotechnol Prog 11:235
43. Tiemann KJ, Gardea-Torresdey JL, Gamez G, Dokken K, Cano-Aguilera I, Renner MW, Furenlid LR (2000) Environ Sci Technol 34:693
44. Gardea-Torresdey JL, Tiemann KJ, Gamez G, Dokken K, Cano-Aguilera I, Furenlid LR, Renner MW (2000) Environ Sci Technol 34:4392
45. US EPA (1998) Source Book on Natural Landscaping for Local Officials, 2nd edn. Northeastern Illinois Planning Commission, Chicago, IL
46. Gatliff EG (1994) Remed Summer 4(3):343
47. Davis LC, Muralidharan N, Visser VP, Chaffin C, Fateley WG, Erickson LE, Hammaker RM (1994) Alfalfa Plants and Associated Microorganisms Promote Biodegradation Rather than Volatilization of Organic Substances from Ground Water. In: Anderson TA, Coats JR (eds) Bioremediation Through Rhizosphere Technology. ACS Symposium Series 563, Washington, DC, p 112
48. Ferro AM, Kennedy J, Doucette W, Nelson S, Jauregui G, McFarland B, Bugbee B (1997) Fate of Benzene in Soils Planted with Alfalfa: Uptake, Volatilization, and Degradation. In: Kruger EL, Anderson TA, Coats JR (eds) Phytoremediation of Soil and Water Contaminants. ACS Symposium Series No. 664, Washington, DC, p 223
49. Jordahl JL, Foster L, Schnoor JL, Alverez PJJ (1997) Environ Toxicol Chem 16:1318
50. Carman EP, Crossman TL, Gatliff EG (1997) Phytoremediation of Fuel Oil Contaminated Soil. In: Leeson A, Alleman BC (eds) Proceedings of the Fourth International In Situ and On-Site Bioremediation, Vol 4(3), Battelle Press, Columbus, OH, p 347
51. Tsao DT (2000) Wildlife Habitat Council Conference on The New Frontier: Brownfield Development and Superfund/RCRA Remediation Projects, Washington, DC
52. Frazar C (2000) The Bioremediation and Phytoremediation of Pesticide-Contaminated Sites. National Network of Environmental Studies Document. U.S. Environmental Protection Agency Solid Waste and Emergency Response
53. Anderson TA, Walton BT (1995) Environ Toxicol Chem 14:2041
54. Baker AJM, Brooks RR, Reeves RD (1998) New Science 177:44
55. Keiffer CH, Ungar IA (1996) Bioremediation of Brine Contaminated Soils, Petroleum Environment Research Forum Final Report #91–18
56. Raskin I, Nanda Kumar PBA, Dushenkov S, Blaylock MJ, Salt D (1994) Phytoremediation – Using Plants to Clean Up Soils and Waters Contaminated with Toxic Metals. In: Emerging Technologies in Hazardous Waste Management VI. ACS Special Symposium, Vol I, Atlanta, GA
57. Baker AJM, McGrath SP, Sidoli MD, Reeves RD (1995) Mining Environ Management 3(3):12
58. Nanda Kumar PBA, Dushenkov V, Motto H, Raskin I (1995) Environ Sci Technol 29:1232

59. Martin HW, Young TR, Kaplan TI, Simon L, Adriano DC (1996) Plant and Soil 182:199
60. LaCoste C, Robinson B, Brooks R, Anderson C, Chiarucci A, Leblanc M (1999) Internat J Phytorem 1(4):327
61. Saxena PK (1999) International Business Communication's Fourth Annual International Conference on Phytoremediation, Toronto, Canada
62. Banuelos GS, Ajwa HA, Terry N, Downey S (1997) Phytoremediation of Selenium-Laden Effluent. In: Leeson A, Alleman BC (eds) Proceedings of the Fourth International In Situ and On-Site Bioremediation Symposium, Vol 4(3). Battelle, Columbus, OH, p 303
63. Ma LQ, Komar KM, Tu C, Zhang W, Cai Y, Kennelley ED (2001) Nature 409:579
64. Entry JA, Vance NC, Hamilton MA, Zabowski D, Watrud LS, Adriano DC (1996) Water Air Soil Pollution 88:167
65. Vasudev D, Ledder T, Dushenkov S, Epstein A, Kumar N, Kapulnik Y, Ensley B, Huddleston G, Cornish J, Raskin I, Sorochinsky B, Ruchko M, Prokhnevsky A, Mikheev A, Grodzinsky D (1996) Removal of Radionuclide Contamination from Water by Metal-Accumulating Terrestrial Plants. In Situ Soil and Sediment Remediation Conference, New Orleans, LA.
66. Negri MC, Hinchman RR, Wozniak JB (2000) Phytoremediation State of the Science Conference, Boston, MA
67. Delvaux B, Kruyts N, Cremers A (2000) Environ Sci Technol 34:1489
68. Briggs GG, Bromilow RH, Evans AA (1982) Pesticide Science 13:495
69. Newman LE, Gordon MP, Heilman P, Cannon DL, Lory E, Miller K, Osgood J, Strand SE (1999) Soil & Groundwater Cleanup Feb/Mar:42
70. Garrison AW, Nzengung VA, Avants JK, Ellington JJ, Jones WJ, Rennels D, Wolfe NL (2000) Environ Sci Technol 34:1663
71. Burken JG, Schnoor JL (1997) Uptake and Fate of Organic Contaminants by Hybrid Poplar Trees. In: American Chemical Society Division of Environmental Chemistry Preprints of Extended Abstracts, Vol 37(1), 213th ACS National Meeting, San Francisco, CA, p 302
72. Zhang Q, Davis LC, Erickson LE (2001) Environ Sci Technol 35:725
73. Heaton ACP, Rugh CL, Wang *N*-J, Meagher RB (1998) J Soil Contam 7(4):497
74. Adler T (1996) Sci News 150:42
75. Viessman W, Lewis GL, Knapp JW (1989) Introduction to Hydrology, 3rd ed, Harper & Row Publishers, New York, NY
76. Kadlec RH, Knight RL (1996) Treatment Wetlands. CRC Press, Boca Raton, FL
77. Lewis-Hutchinson S, Weaver J (2000) Phytoremediation State of the Science Conference, Boston, MA.
78. Baker JM, Van Bavel CHM (1987) Plant Cell Environ 10:777
79. Granier A, Huc R, Barigah ST (1996) Agric Forest Meteorol 78:19
80. Hirsch S, Compton H (2000) Phytoremediation State of the Science Conference, Boston, MA
81. Ansley RJ, Dugas WA, Heuer ML, Trevino BA (1994) J Exp Bot 45:847
82. Messina MG, Duncan JE (1993) Agric Water Management 24:265
83. Dugas WA, Wallace JS, Allen SJ, Roberts JM (1993) Agric Forest Meteorol 64:47
84. Hinckley TM, Brooks JR, Cermak J, Ceulemans R, Kucera J, Meinzer FC, Roberts DA (1994) Tree Physiol 14:1005
85. Wiltshire JJJ, Unsworth MH, Wright CJ (1994) New Phytol 127:349
86. Dugas WA, Mayeux HS Jr (1991) J Range Management 44(2):161
87. Landemeyer JE (2000) Phytoremediation State of the Science Conference, Boston, MA
88. Quinn JJ (2000) Phytoremediation State of the Science Conference, Boston, MA
89. Steinberg SL, van Bavel CHM, McFarland MJ (1989) J Amer Soc Hort Sci 114(3):466
90. Heilmann JL, Ham JM (1990) HortScience 25(4):465
91. Steinberg SL, van Bavel CHM, McFarland MJ (1990) Agron J 82:851
92. Steinberg SL, Zajicek JM, McFarland MJ (1991) J Amer Soc Hort Sci 116(3):460
93. Grime VL, Morison JIL, Simmonds LP (1995) Agric Forest Meteorol 74:1
94. Steiner JL, Howell TA, Schneider AD (1991) Agron J 83:240
95. Allen RG, Jensen ME, Wright JL, Burman RD (1989) Agron J 81:650

96. Hinchman RR, Negri MC, Gatliff EG (1997) Phytoremediation Using Green Plants to Clean Up Contaminated Soil, Groundwater, and Wastewater. US Department of Energy, Energy Efficient and Renewable Energy Contract W-31 – 109-Eng-38
97. Chappell J (1998) Phytoremediation of TCE in Groundwater Using *Populus*. US EPA Technology Innovation Office National Network of Environmental Management Studies Fellowship
98. Harvey G (1998) Third International Conference on Phytoremediation, Houston, TX
99. Chard JK, Orchard BJ, Pajak CJ, Doucette WJ, Bugbee B (1998) Conference on Hazardous Waste Research, Snow Bird, UT
100. Ferro AM (1998) Third International Conference on Phytoremediation, Houston, TX
101. USDA-NRCS (2001) The PLANTS Database, Ver 3.1 (http://plants.usda.gov), National Plant Data Center, Baton Rouge, LA
102. Wraith JM, Baker JM (1991) Soil Sci Soc Am J 55:928
103. Vetterlein D, Marschner H, Horn R (1993) Plant and Soil 149:263
104. Weand BL, Hauser VL (1997) J Environ Prot, November, p 40
105. NRC (1993) Vetiver Grass: A Thin Green Line Against Erosion. National Research Council, National Academy Press, Washington
106. Kulakow P (2000) Annual Report of the RTDF Phytoremediation Action Team – TPH Subgroup Cooperative Field Trials, Great Plains/Rocky Mountain Hazardous Substance Research Center, Kansas State University, Manhattan, KS
107. Blaylock MJ, Salt DE, Dushenkov S, Zakharova O, Gussman C, Kapulnik Y, Ensley BE, Raskin I (1997) Environ Sci Technol 31:860
108. Dodge CJ, Francis AJ (1997) Environ Sci Technol 31:3062
109. Riesen TK, Bruner I (1996) Environ Pollut 93(1):1
110. Dushenkov S, Kapulnik Y, Blaylock M, Sorochisky B, Raskin I, Ensley B (1997a) Studies Environ Sci 66:563
111. API (1998) The Use of Treatment Wetlands for Petroleum Industry Effluents. America Petroleum Institute Health and Environmental Sciences Department Publication #4672
112. Knight RL, Kadlec RH, Ohlendorf HM (1999) Environ Sci Technol 33(7):973
113. Bhadra R, Spanggord RJ, Wayment DG, Hughes JB, Shanks JV (1999) Oxidative Metabolism of 2,4,6-Trinitrotoluene in Aquatic Phytoremediation Systems of *Myriophyllum aquaticum*. In: Means JL, Hinchee RE (eds) Wetlands and Remediation, Battelle, Columbus, OH, p 127
114. Ferro AM, Kennedy J, Sorenson D (1999) Wetland and Remediation: An International Conference, Salt Lake City, UT
115. Medina VF, Rivera R, Larson SL, McCutcheon, SC (1998) Soil Groundwater Cleanup Feb/Mar, p 19
116. Baker JG, Revel RD (1999) Vegetation Establishment in an Urban Constructed Wetland: Results from a Canadian Pilot Project. In: Means JL, Hinchee RE (eds) Wetlands and Remediation. Battelle, Columbus, OH, p 59
117. Wilkinson F, Beckett PJ, St-Germain P (1999) Mine Water Environment, IMWA Congress, Sevilla, Spain, p 591
118. USDA (2000) Forest Nursery Notes, US Department of Agriculture, Forestry Service Document # R6-CP-TP-01 – 00
119. Gatliff EG (2000) Phytoremediation State of the Science Conference, Boston, MA
120. Dushenkov S, Nanda Kumar PBA, Motto H, Raskin I (1995) Environ Sci Technol 29(5):1239
121. Dushenkov S, Vasudev D, Kapulnik Y, Gleba D, Fleischer D, Ting KC, Ensley B (1997b) Environ Sci Technol 31(12):3468

Received: November 2001

Soil Plant Microbe Interactions in Phytoremediation

R. Karthikeyan[1] · P. A. Kulakow[2]

[1] 147 Seaton Hall, Kansas State University – Department of Biological and Agricultural Engineering, Manhattan, KS 66506, USA. *E-mail: karthi@ksu.edu*

[2] 2004 Throckmorton Plant Center, Kansas State University – Department of Agronomy, Manhattan, KS 66506–5501, USA. *E-mail: kulakow@ksu.edu*

Use of vegetation in remediation of soil and groundwater contaminated with organic materials is a promising, cost-effective alternative to the more established treatment methods used at hazardous waste sites. Plants can transpire groundwater and lower the concentrations of organic contaminants in soils and groundwater. The evapotranspirational activity of vegetation acts as a natural pump-and-treatment system. Plants have shown the capacity to absorb, uptake, and convert organic contaminants to less toxic metabolites in laboratory and field studies. Vegetation also plays a significant role in bioremediation. This is because plants stimulate the degradation of organic compounds in the rhizosphere by the release of root exudates and enzymes. Success of any plant-based remediation system depends on the interaction of plants with the surrounding soil medium and the contaminant. Knowing the fate of an organic contaminant in the soil can help determine the persistence of the contaminant in the terrestrial environment and ultimately the success of any remediation method. Also, an understanding is needed of soil-plant-microbe interactions that determine the fate of organic contaminants in the soil-plant ecosystem. This paper presents an overview of the subsurface environment and fate and transport processes of organic contaminants as affected by soil-plant-microbe interactions.

Keywords. Bioremediation, Organic contaminants, Rhizosphere, Transport

Advances in Biochemical Engineering/
Biotechnology, Vol. 78
Managing Editor: T. Scheper

List of Symbols

C_a	contaminant concentration in soil-air phase (ML^{-3})
C_s	contaminant concentration in soil-solid phase (MM^{-1})
C_T	total contaminant concentration in different phases (ML^{-3})
C_w	contaminant concentration in soil-water phase (ML^{-3})
D	soil aqueous phase diffusion coefficient – lumped parameter (L^2T^{-1})
D_z	soil aqueous phase diffusion coefficient (L^2T^{-1})
D_{za}	soil gas phase diffusion coefficient (L^2T^{-1})
f_{oc}	fraction of organic carbon in soil (MM^{-1})
H	Henry's constant ($ML^{-3}M^{-1}L^3$)
J_c	rate of contaminant mass flow per unit area ($ML^{-2}T^{-1}$)
$J_{c,adv}$	rate of contaminant advection per unit area ($ML^{-2}T^{-1}$)
$J_{c,diff}$	rate of contaminant diffusion per unit area ($ML^{-2}T^{-1}$)
k_d	soil-water partitioning coefficient (L^3M^{-1})
k_{oc}	soil organic carbon mass fraction (L^3M^{-1})
q_z	velocity of water moving in z direction – lumped parameter (LT^{-1})
r	rate of contaminant mass disappearance per unit volume ($ML^{-3}T^{-1}$)
v_z	velocity of water moving in z direction (LT^{-1})
β	overall first-order decay coefficient for contaminant (T^{-1})
γ	overall first-order plant uptake constant for contaminant (T^{-1})
Δx	arbitrarily small length in x direction (L)
Δy	arbitrarily small length in y direction (L)
Δz	arbitrarily small length in z direction (L)
θ_a	volumetric air content (L^3L^{-3})
θ_w	volumetric water content (L^3L^{-3})
κ	retardation coefficient
ϱ	soil bulk density (ML^{-3})
μ_w	first-order decay coefficient for contaminant in aqueous phase (T^{-1})
μ_s	first-order decay coefficient for contaminant in sorbed phase (T^{-1})
μ_{wp}	first-order plant uptake coefficient in aqueous phase (T^{-1})

1
Introduction

The decontamination of soils contaminated with various anthropogenic chemicals is a global problem that consumes considerable economic resources [1, 2]. Over the past two decades, pollution prevention and the cleanup of contaminated soils have become a worldwide environmental priority. Several suitable technologies have been developed to achieve the goal of soil cleanup including the use of natural organisms ranging from soil microorganisms up to higher plants.

The complex xenobiotic nature of most synthetic chemicals requires the interaction of microbial communities at the molecular and ecological levels to achieve biotransformation [3]. Recently, plants have been studied for their ability to facilitate biotransformation processes [4–23]. Bioremediation in the presence of vegetation, often known as phytoremediation, is an inexpensive cleanup method for treating contaminated sites impacted with organic and/or inorganic constituents [5, 24]. In addition to the properties, functions, concentrations, and distribution of contaminants, successful phytoremediation of soil depends on several interrelated factors including the soil conditions, transport processes, microbial characteristics, and type of vegetation.

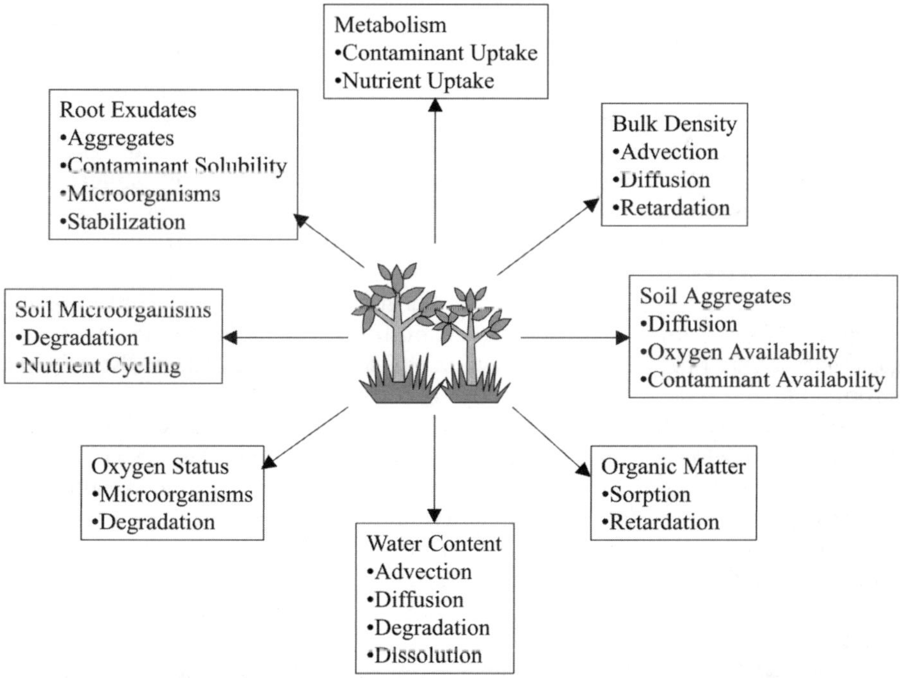

Fig. 1. Effects of plants on soil properties that ultimately affect the fate and transport processes in the subsurface

Soil, plant, and microbe relationships are interdependent and complex. A change in one component typically affects the other two. Plants have a number of important influences on both soil and microbial interactions that ultimately affect the effectiveness of phytoremediation. Plants alter the soil physical, chemical, and biological properties, as well as alter the fate and transport of a contaminant in the subsurface. Figure 1 summarizes the effects of plants on soil properties that ultimately affect fate and transport processes in the subsurface.

2
Characteristics of the Subsurface Environment

Soil is a complex system made up of a heterogeneous mixture of solid, liquid, and gaseous components. The solid phase is composed of a mineral portion and an organic fraction, containing particles of varying physical and chemical compositions and properties [25]. The organic fraction is complex, containing a diverse population of living, active organisms as well as plant and animal residues in various stages of decomposition [26]. The liquid phase of soil consists essentially of soil water, which fills part or all of the void spaces between the solid particles. Soil water contains solutes that may have been dissolved from the soil mineral phase or may have entered through the soil surface by natural or anthropogenic means. Soil water is held by forces in the soil matrix and varies significantly in mobility depending on its location. The gaseous phase of soil occupies empty void spaces that are not filled with water and other liquids. This phase may have gases that differ considerably from that of the air above the soil surface and may undergo dramatic changes in a short time period. Relationships among the constituents of the three phases are affected by the chemical and physical properties of each component as well as by temperature, pressure, and light [25].

Soils are open systems that undergo continuous biological and chemical transformations that link them physically with the atmosphere and hydrosphere [27]. One mode of interaction is the presence of plants grown in the soil. While soils provide the supporting medium and a nutrient source for plant life, plants build up the organic matter content in the soil matrix and support the microbial community. Furthermore, plants also take up and transpire soil water into the atmosphere. This changes the hydrologic conditions which dictate the liquid and gaseous phases in the soil matrix. Each of these interactions affects the soil environment.

Plants play a significant role in altering soil physical, chemical, and biological properties. The effect of plants on the soil environment can alter fate and transport processes of a contaminant. Plants can alter the mobility of a contaminant by changing the soil water content. Plants can also release root exudates, which can lead to a change in the soil pH. Altering soil pH could affect trace element movement and cycling in soils. Furthermore, plants can increase the number of bacteria around root surfaces, which might stimulate biodegradation of organic compounds.

While plant roots have an important influence on soil structure, the mechanisms by which roots alter soil structure are obscure. Plant roots impart structural changes that cannot be attributed only to the production of decomposable plant residue [28]. As discussed in detail below, plants also affect the soil bulk density, organic matter, aggregation, water content, and oxygen availability. Furthermore, all of these properties ultimately affect the fate and transport of organic contaminants in soil.

2.1 Bulk Density

The bulk density of soil is defined as the dry weight of soil per unit volume. This soil characteristic can be changed through several natural and man-made processes. In general, soil compaction results in decreased volume and hence increased bulk density. Changes in soil compaction resulting from the use of heavy machinery can cause a decrease in the volume of the soil. Similarly, soil volume can be altered by forces exerted by natural sources such as the development of plant roots. When roots develop in an unsaturated soil, the soil water that is present is displaced from void spaces and the contact area between adjacent soil particles increases. This increase depends on the degree of deformation of the soil particles by plant roots. However, the deformation is usually assumed to be elastic in nature and the soil particles will rebound upon the death of the roots. Therefore, soil compaction depends on the arrangement of soil particles and the degree of load applied by roots.

Bulk density alters the soil structure in terms of the volume, size, and shape of the void spaces. Therefore, water content and mass transport pathways in soils are significantly affected as well. An increase in bulk density, at constant gravimetric water content, increases the volumetric water content and hence, contaminant diffusion. In addition, water percolating through the soil can carry dissolved contaminants through mass flow. By changing bulk density, the mass flow of contaminants can be altered by creating gradients of total pressure. Similarly, dissolved gases in the percolating water derived either from the atmosphere or from volatile contaminants can be carried as well. However, gas transport within the soil, as well as between the soil and the atmosphere, occurs mainly in air-filled soil pores. Increasing bulk density reduces soil porosity, which can stop or limit the transfer of volatile contaminants in the gaseous phase. However, as gaseous diffusion is reduced, the pore size distribution becomes altered as well, frequently resulting in the loss of the larger pores where the majority of mass transfer occurs. Finally, the content and transport of heat in soils are also affected by changes in bulk density. This could affect the soil temperature. Since the degradation of several contaminants are temperature dependent, these changes in soil temperature could alter the degradation kinetics.

Plants transport oxygen from the atmosphere to the root zone. During this process, pockets of air can be created around the roots. This will loosen the soil particles, increasing the volume of soil per given weight. This results in a lower bulk density. Similarly, when plant roots decompose, the resulting space leads to increased soil volume and decreased bulk density as well. Furthermore, the permeability of a soil increases as water is able to flow in the channels created by the decomposed roots. This was demonstrated in a synthetic sandy loam where the roots initially caused a reduction of soil pore space due to compaction by the roots, but after the root residue decomposed, permeability increased and density decreased [29].

As bulk density is reduced at constant gravimetric water content, the volumetric water content is decreased as well, leading to a decrease in the aqueous phase diffusion of a soluble contaminant. At the same time, gas phase diffusion is increased as additional pore spaces become available for increased gas and oxygen transport to occur. This might facilitate the volatilization of some contaminants in soil. Also, more oxygen will be available in the gaseous phase for aerobic microorganisms. This will ultimately increase the aerobic degradation of some organic compounds.

Although plant growth may have long term effects of decreasing soil bulk density, compacted soil with relatively high bulk density can severely limit plant root development and plant growth. This can be a problem on industrial sites with poor soil structure and compaction by the use of heavy equipment [30]. Successful establishment of plants at contaminated sites depends on soil management and tillage operations that minimize compaction.

2.2 Soil Aggregation

Soil aggregation is a controlling factor for microbial activity and soil organic matter turnover. Aggregate formation is initiated when soil microflora and plant roots produce fibrils, filaments, and polysaccharides that combine with the soil clay fraction to form organomineral complexes. Other physical forces such as drying, shrink-swell, freeze-thaw, compaction, animal movement, and root growth cause the soil to form into aggregates [31–33]. Clays are basic to aggregate formation. Particles involved in aggregate formation also include organic molecules measurable in nanometers, microorganisms and silt particles measurable in micrometers, and small root fibers measurable in millimeters [34]. Aggregates vary greatly in size. Two broad categories, micro- and macro-aggregates, are commonly used. Micro-aggregates are less than 250 μm in diameter while macro-aggregates are more than 250 μm.

Soil aggregation increases the permeability of soil. Increased soil aggregation facilitates water and nutrient transport by inducing macro-pore flow [35–38]. This may facilitate the degradation of contaminants by increasing the transport of alternate electron acceptors where the availability of acceptors is generally low or depleted. However, the pore spaces within aggregates may be closed to bulk flow. The fate of microorganisms and contaminants contained inside an aggregate aggregate will be completely different from that of contaminants located outside of an aggregate. Contaminants contained within an aggregate will be affected primarily by oxygen transfer limitations. Thus, the interior pores of soil aggregates could be anaerobic with the contaminant trapped inside. Therefore, the contaminant would only be subjected to anaerobic biodegradation, which is not as efficient as aerobic biodegradation for many contaminants [39–44]. In addition, aggregates have potential adsorption sites for soil enzymes such as ureases and proteases. These enzymes, whether microbial or plant-produced, would not be able to interact with contaminants contained within a soil aggregate. Consequently, these contaminants would be broken down at a much slower rate than would contaminants located outside of an aggregate. Contaminants inside the aggregate can persist for years and may not be bioavailable to the microorganisms.

An example of how aggregation can affect the fate of a contaminant is trichloroethylene (TCE), a toxic compound that can be anaerobically biotransformed to vinyl chloride [45–48]. If TCE is trapped inside an aggregate, it could be degraded to vinyl chloride in the anaerobic pore environment. However, vinyl chloride, another toxic compound, can only be completely degraded to its basic components under aerobic conditions. Therefore, inside the closed aggregate, vinyl chloride could build up, leading to a continued threat if the aggregates are not broken up.

2.3 Organic Matter

Organic matter is an extremely important component of soil. Small changes in organic matter content make large differences in the physical characteristics of a soil including sorption capacity, water retention, and structure formation. It also has a dramatic effect on chemical properties such as cation exchange capacity and chelation of metals and other ions. In terms of mass, microorganisms generally contribute very little to the overall soil organic matter content. However, the actions of these organisms are largely responsible for the fate of organic matter and other contaminants in soil. Most soil organic matter is derived from plants either as exudates or plant residues. Plant organic matter is composed of proteins, fats, sugars, and an assorted mixture of polymers and polymeric materials. Nitrogen-containing compounds, including complex polypeptides and nucleic acids, are rapidly decomposed by microorganisms. Oils and fats are more slowly decomposed, in part because they are insoluble in water. Polymeric compounds made up of sugars are resistant to decomposition. Fats, oils, and polymer sugars are low in nitrogen, which limits their decomposition rate [49].

Soil organic matter plays a crucial role in the fate and transport of many organic contaminants. Hydrocarbons tend to partition to soil organic matter, and once sorbed, their degradation may be dependent on the rate at which these chemicals desorb into the aqueous phase [26, 50–54]. Hence, soil organic matter plays an important role in hydrocarbon degradation. The presence of vegetation may increase the soil organic matter content by releasing root exudates. Some hydrocarbons may become part of the soil humus via microbial biomass or by more direct incorporation [55–62].

Exudates released by microorganisms and plant roots are important to soil pH, nutrient cycling of trace elements, and the movement of trace metals. It is also noted that plant roots exude several organic acids commonly found in soils [26]. The environment around plant roots has been studied from different perspectives including ex planta enzymatic effects [63–65] and plant associated microflora effects [7, 55, 56, 66–68]. The levels of enzymatic production are dependent on the plant species. Furthermore, there also may be differences in the rate and timing (age, season, stress induced, etc.) of release of the enzymes that have yet to be thoroughly investigated.

Several studies observed that plant enzymes can metabolize a wide variety of xenobiotic pollutants. These enzymes can be found both internally or externally to the plant system, depending on the specific enzyme. A greater variety of active plant enzymes have been discovered in sediments far from their plant source, and some of the highly characterized plant enzymes that are exuded by the plant root include certain oxidoreductases, peroxidases, and laccases [65].

2.4 Water Content

Soil water extraction by plant roots significantly alters the soil water content which, in turn, alters microbial processes, plant growth, and the fate and transport of contaminants [21, 69–71]. The two most important characteristics of the soil water phase are the amount of water in a specified quantity of soil and the force holding the water in the soil matrix. The amount of water in the soil influences many processes, including gas exchange with the atmosphere, diffusion of contaminants and nutrients to plant roots, diffusion of contaminants within the soil matrix, soil temperature, and the speed with which contaminants move through the root zone during rainfall or irrigation. The forces retaining water in the soil matrix also affect many soil processes including the efficiency of plant water extraction, the amount of leaching occurring under gravity, and the extent of upward capillary movement of water and contaminants against gravity. Also, the soil water regime has many interacting factors that singly, or in combination, directly or indirectly exert a selective pressure on living organisms [72, 73].

Soil-water relationships can significantly affect the decomposition of organic compounds and the rate of organic matter turnover [74]. As the water content of soils decreases due to plant uptake, the percentage of air-filled pores increases, resulting in an increased oxygen tension in the soil. As a result, the soil becomes more aerobic, which would increase the degradation rates of some contaminants by stimulating microbial activity. However, this microbial activity is balanced in relatively dry soils since these do not provide enough water required by the microorganisms [75, 76]. Soil microorganisms existing in a microscopic cavity within soil will remain active only when nutrients and water are available. Continued activity involves either movement of the bacteria to a new substrate or the movement of the substrate to the bacteria by plant growth and nutrient diffusion [72]. The optimum water content for decomposition is when 60% of the soil pores are filled with water [72, 74, 76, 77]. When the moisture content increases above 60%, decomposition can decrease or cease altogether. This is due to limited oxygen availability as well as the decreased overall microbial activity.

Carbon, nitrogen, and other nutrient cycling involving soil microorganisms depend on soil water content as well. Specifically, the diffusion of oxygen, nitrate, and dissolved carbon are all affected by water content [78]. In many soils, a plant canopy as small as 15 cm in height can stop the leaching of nitrate [77]. This is purely by lowering the soil water content and not attributable to nitrate uptake. Furthermore, aerobic nitrifiers and anaerobic denitrifiers are very sensitive to water content [79]. In general, the aerobic processes occur when 20 to 60% of the pores are filled with water while anaerobic processes occur when more than 60% of the pores are filled. Below 20%, insufficient water is present to act as an adequate medium to support the microorganisms. Also, above 60%, nitrification decreases rapidly as oxygen becomes limiting. Conversely, denitrification increases to above 60% and peaks when close to 100% of the pores are filled with water. In either case, rates of nitrification and denitrification can be enhanced by the carbon supplied by plants.

Growth behavior of plant roots is also closely related to soil water content. Specifically, a direct relationship exists between the soil water content and the mean daily water uptake rate by vegetation [70, 71]. In a phytoremediation study, a high mean daily water uptake rate by the plants induced a relatively large water potential gradient. This resulted in a moisture deficiency in the upper soil layers [21]. This moisture deficiency may greatly limit the ability of plants to take up water and may encourage the growth of deeper roots in order to absorb moisture from lower soil layers [27].

2.5 Oxygen Availability

Living organisms, particularly soil organisms can obtain energy from the oxidation of reduced materials. Regardless of the availability of carbon, microorganisms must utilize an electron acceptor in order for degradation to occur. In many cases, oxygen acts as the terminal electron acceptor for organic contaminants [1, 77, 80–84]. Furthermore, several studies have shown that many hydrocarbons absolutely require oxygen in order for significant biodegradation to occur [85, 86]. In many cases, the concentration of available oxygen is the limiting factor in biodegradation.

In the absence of oxygen, oxidized forms of nitrogen, manganese, and iron can serve as electron acceptors [87]. Given that the solubility of oxygen in water is only 8.6 mg/L (at 25° C and 1 bar) [88], depletion of available oxygen will be rapid when the demand is high. This situation is further magnified by the fact that the diffusion of oxygen in water is about 10^{-4} times the diffusion of oxygen in air. The corresponding oxygen diffusion coefficients for water versus air are 2.5×10^{-5} cm^2 s^{-1} and 0.189 cm^2 s^{-1}, respectively [89]. Therefore, the aerobic biodegradation rate in the soil matrix often becomes a function of the oxygen resupply to those regions.

The presence or absence of oxygen depends on several factors such as soil water content and bulk density. The oxygen present in the soil depends on the total amount of air-filled pore spaces, the size of the pores, the rate of oxygen consumption, and the geometric distribution of the aerobic microorganisms in the soil layer. Higher amounts and larger air-filled pore spaces insure the existence of high oxygen reserves and also allow a rapid replacement of oxygen by diffusion to occur. The elimination of air-filled pore spaces by waterlogging or compaction, for example, reduces oxygen reserves to the small amounts dissolved in the soil solution. In fine-textured, heavy clay soils, the small size of the pores also impedes oxygen diffusion. Furthermore, the presence of metabolizable organic substrates, including hydrocarbons, tends to deplete the oxygen reserve over time, especially if the resupply rate is limited. In addition, the rate of oxygen diffusion to deeper soil layers decreases as the thickness of an overlying layer containing a high aerobic activity increases. Therefore, the upper few centimeters of a contaminated soil may remain aerobic while the deeper layers can quickly become anaerobic. However, plant roots may improve the diffusion of oxygen to deeper soil layers by providing root channels for airflow. This may occur as roots decay and dessicate or by decreasing the soil bulk density.

2.6 Soil Microorganisms

Soil provides habitat for a wide range of organisms ranging from microbes to higher species such as plants. Soil biological properties include the activity and diversity of soil microorganisms. Bacteria are the smallest and most numerous of the organisms in soil with more than 400 named genera and an estimated 10^4 species [77]. On a gram dry-weight basis, microbial numbers in soil often exceed 10^8 for bacteria, 10^6 for actinomycetes, and 10^5 for fungi. Microbial biomass can exceed 500 mg carbon per kg soil [90]. However, microorganisms only comprise less than 3% of the soil organic carbon and take up only 0.001% of the total soil volume [91]. The density of microbial populations varies significantly both vertically and laterally in the soil profile. Dispersion of microbial biomass within the profile is related to the distance from the soil surface due to oxygen diffusion, as well as the placement and availability of organic materials as potential food sources [66].

Most bacteria in the subsurface have a tendency to attach to solid particles and form a biofilm which is rich in substrates desirable by microorganisms [92]. Because of the abundant release of root exudates and mucilages, a rapid buildup of a biofilm generally occurs within 50 µm from the root surface. In this vicinity, the microbial population size can be 50–100 times greater than in non-vegetated soil. Both quantitative and qualitative measurements show that bacterial coverage of root surfaces usually ranges from 5 to 10% [77]. Single bacteria are often associated with pits in root cell walls while clusters of bacteria inhabit the void spaces between cells. This increased population significantly increases the microbial activity and, hence, the biodegradation rates.

The term "rhizosphere" was first introduced in 1904 to describe the specific interaction between bacteria and the roots of legumes. The term has since been refined to regard the rhizosphere as the soil region under the immediate influence of plant roots where there is an increased microbial population (see Table 1). The rhizosphere is known to harbor a great diversity of forms with proportionately more denitrifiers and gram negative bacteria (*Pseudomonas*, *Achromobacter*) rather than gram positive bacteria and other forms [77]. Furthermore, the rhizosphere microbial community typically is different than that of the bulk soil as a result of differences in chemical and physical properties.

Table 1. Number of bacteria at increasing distance from the root surface [77]

Distance from surface (mm)	Estimated frequency (10^9 cells cm^{-3})	Morphological types distinguishable
0–1	120	11
1–5	96	12
5–10	41	5
10–15	34	2
15–20	13	2

Most members of the soil biota are organotrophs. The major source of carbon for these soil organisms are plant roots and organic residues contributed during and following plant growth. Plants release nutrients such as amino acids, simple sugars, carbohydrates, and enzymes into the soil that are potential substrates for the microorganisms. Root exudates usually are low molecular weight substances that leak from plant cells into the soil either through the spaces between cells or directly from the epidermal cell walls. Root secretions and plant hormones include low molecular weight compounds and high molecular weight mucilages, both of which are released as a result of metabolic processes. Root turnover is a major mechanism that adds organic carbon to the soil matrix. This organic carbon is a potential energy source for microorganisms. Analyses of organic compounds found on, in, or near roots reveal a wide variety of aliphatic, amino, and aromatic acids as well as amides, sugars, and amino sugars. In addition to such soluble and diffusible compounds, there are several insoluble compounds occurring in the roots such as cellulose, lignin, and protein. These insoluble compounds are lost from roots by cell exfoliation and root pruning. Some of the more complex aromatic acids and enzymes occur only in the root zones of specific species.

During active growth, microorganisms need 30 parts of carbon for each part of nitrogen they utilize [77, 93]. However, microorganisms can rapidly deplete a soil of the available carbon and nitrogen derived from plant decomposition, root exudates, and microbial decomposition of soil organic matter [94, 95]. In addition, phosphorus and other inorganic nutrients in the aqueous phase are essential for microbial growth and biodegradation. Therefore, when considering biodegradation processes in soil environments, the relationship between the active population and available nutrient source is critical.

Microorganisms can be considered as biochemical reactors that drive the biodegradation process. This idea is a logical extension of many studies showing that the turnover of soil organic matter and nutrient cycling processes are microbially driven [94]. Physiochemical factors that determine the suitability of the soil environment as microbial habitats, and thus, sites for biodegradation are given in Table 2. Bacteria are the principal microbiota that mineralize many organic contaminants with numerous studies documenting their degradation capabilities [82, 96–112]. The predominant microorganisms are *Pseudomonas, Mycobacterium, Actinobacter, Flavobacterium, Arthrobacter, Bacillus,* and *Nocardia* [11, 83, 113].

Table 2. Physical and chemical characteristics of soils as microbial habitats [98]

Parameters	Soils
Expanse	21% of the global surface
Temperature	−40 °C to 65 °C
pH	2.5 to 11.0; most soils are slightly acidic
Water Potential	−100 MPa to −0.001 MPa
Salinity	From near zero to salt saturation, but low in the majority of soils
Oxygen Availability	Adequately oxygenated except when water logged
Inorganic Nutrients	N and P are frequently limiting
Organic Matter	Relatively abundant but consisting mostly of refractory humic substances
Attachment Surfaces	Great abundance of inorganic and organic surfaces

3 Contaminant Fate and Transport Processes

The fate and transport of a contaminant in the soil matrix depends on soil physical, chemical, and biological properties. The ability of the contaminant to interact in this environment determines its persistence while biotic fate processes, such as biodegradation determine the bioremediation potential. The subsurface fate of a contaminant is largely determined by mass transport processes (advection, diffusion, and dispersion) and mass transfer processes (dissolution, hydrolysis, sorption, volatilization, and biological processes). In some cases, mass transfer processes may also include oxidation-reduction (redox) reactions, acid-base reactions, complexation, precipitation, and radioactive decay [114].

3.1 Dissolution and Hydrolysis

Dissolution refers to the solubilization of a compound. Water is considered to be a universal solvent since many chemicals in solid, gas, and liquid phases can dissolve in water. It is important to distinguish between organic contaminants that are insoluble or immiscible in water from those that are partially or completely soluble in water. An organic contaminant that exists as a separate, immiscible phase when in contact with water and/or air is known as a non-aqueous phase liquid (NAPL). An organic contaminant in physical contact with groundwater may partially or completely dissolve (solubilize or partition) into the aqueous phase. The solubility of an organic compound is the equilibrium concentration of the compound in water at a specified temperature and pressure. For all practical purposes, solubility represents the maximum concentration of a compound in water. Several parameters affecting solubility include chemical activity of a compound, temperature, pH, cosolvents, dissolved organic matter, and dissolved inorganic compounds (salinity).

The solubility of organic compounds is highly variable. In general, charged or polar organic compounds or those containing oxygen or nitrogen are most soluble. Usually, solubility is proportional to the mass of the organic compound.

The octanol/water partition coefficient (K_{ow}) is commonly used to indicate the solubility of an organic compound in water. This parameter is a dimensionless constant which expresses partitioning of the organic compound between octanol and water. The higher the K_{ow}, the lower the water solubility of the organic compound.

Hydrolysis is a mass transfer process that results from the substitution of functional groups between an organic compound and water. The introduction of hydroxyl groups into an organic molecule may make the reaction product more soluble and amenable for biodegradation or plant uptake. Therefore, hydrolysis may be an important mass transfer mechanism in a remediation effort since it may transform an organic compound, originally recalcitrant to biodegradation, to a degradable compound.

3.2 Advection, Diffusion, and Dispersion

Advection is the process by which a contaminant moves with soil pore water. Therefore, it is related to the velocity of water flow and is directly affected by changes in soil water content. Advective transport decreases with a reduction in soil water content because of the decrease in the effective hydraulic conductivity of the contaminant in unsaturated soils [115]. Water-soluble and water-insoluble contaminants differ in the way their transport takes place relative to soil pore water. The transport of soluble contaminants is closely related to the flow of pore water while the transport of immiscible contaminants are governed by a host of pore-scale mechanisms not related to the flow of water.

Another major transport process dictating the fate of organic contaminants is diffusion, which occurs due to contaminant concentration gradients in a system. The diffusion of chemicals in soils is typically grouped with another transport mechanism known as mechanical dispersion, which is the effect of advective velocities causing a mixing of the chemical in soil.

3.3 Sorption

Sorption is the phase transfer process describing the movement of contaminants from a liquid to a solid phase and is a dominant process affecting almost all dissolved contaminants in groundwater. Sorption is defined as the interaction of a contaminant with a solid surface [116–118]. In general, sorption includes adsorption, absorption, and ion exchange. Adsorption is the process of attraction of chemicals to the surface while absorption is the process whereby the chemical is incorporated in the interior of the solid. Ion exchange occurs because of the substitution of cations in solids, and can therefore be considered a special case of absorption. In soil or aquifer material, an organic contaminant will partition onto the solid phase material. The primary pathway in which this process occurs is through the water phase. Numerous parameters affect sorption including solubility, polarity, ionic charge, pH, redox potential, and the octanol/water partition coefficient [117]. Desorption of the contaminant (mass transfer of contaminant from the solid phase to the water phase) is often a rate-limiting step and is partially responsible for the tailing effect commonly observed in groundwater pump-and-treatment systems.

3.4 Volatilization

Volatilization involves the mass transfer of a compound from a liquid or solid phase into the gaseous phase. It is controlled by vapor pressure, which is the pressure of the gas in equilibrium with the liquid or solid at a given temperature. Vapor pressure indicates the solubility of a compound in gas and is therefore an indicator of the tendency of the compound to evaporate. This process is governed by two laws, Henry's law and Raoults' law. Henry's law describes the volatilization or partitioning of an organic compound between the aqueous and gaseous phases. According to this law, the tendency of solute molecules to escape from the water phase into the air phase is proportional to the concentration in water. This relationship assumes local equilibrium between water and air. It is useful for estimating the potential an organic chemical has to move from water to air. It can also be used to estimate the transport of hydrocarbon vapors from air back into the water phase [119].

Volatilization can also occur from a non-aqueous phase. For these concentrated solutions of organic compounds, volatilization is described best by Raoult's law, which states that the vapor pressure over a solution is equal to the product of the mole fraction of the solute and the vapor pressure of the pure-phase liquid. In simple terms, this is a measure of the tendency of molecules to escape from the non-aqueous phase to the gaseous phase. Henry's law and Raoult's law are useful for describing the partitioning of organic compounds between fluid and gaseous phases when the system is at equilibrium [120].

3.5 Biodegradation

Microorganisms act as catalysts for the mass transfer processes described earlier, and in some cases, act as individual bioreactors to degrade contaminants. An organic compound can be biodegraded by oxidation or reduction, by the removal of a functional group, or by partial degradation of the core structure. Mineralization is a process where the original organic molecules are degraded to produce CO_2, microbial biomass, and inorganic compounds containing N, P, and S [121]. The following equation for heterotrophic microbial growth defines the primary nutritional requirements for aerobic biodegradation that occurs in the root zone of plants [7]. Under anaerobic conditions, intermediate materials such as fermentation products or CH_4 will be produced instead.

$$CH_mO_l + aNH_3 + bO_2 \rightarrow y_cCH_pO_nN_q + zCH_rO_sN_t + cH_2O + dCO_2 \tag{1}$$

Many organic contaminants are amenable to these biological degradation processes in the aqueous phase. However, microbial degradation is usually substrate limited. It may be limited by nitrogen, phosphorus, or trace minerals. Under saturated conditions, biodegradation may be oxygen limited. As mentioned earlier, plants support microbial communities in the rhizosphere by providing microorganisms with nutrients via root exudation. Furthermore, nitrogen may be provided from decaying roots or from root nodules by nitrogen fixation. Oxygen may be supplied directly through the roots or by diffusion through root channels. Ultimately, plants can improve the kinetics or rate of microbial degradation.

Bioremediation is the application, stimulation, or enhancement of microorganisms to decontaminate soils and groundwater. Considerable research has focused on evaluating the aerobic and anaerobic biodegradation and transformation processes that play an important role in the ultimate fate of a contaminant in the subsurface [44, 83, 98, 122, 123]. Furthermore, these contaminant reduction pathways can be stimulated under naturally occurring conditions and through actively engineered remediation processes, including phytoremediation. The possibility of microbial degradation of xenobiotics in and near the root zone is promising. Many studies demonstrate the degradation of organic contaminants in the root zone [4, 10, 17, 20, 24, 67, 124–129].

3.6 Plant Uptake

Some contaminants in soil and groundwater may be taken up by plant roots. One important plant uptake mechanism for some compounds is the potential absorption from the gas phase by the roots or leaves. However, the movement of contaminants in the subsurface is primarily affected by the flow of water to the roots. Plants transport water through the xylem and solutes through the phloem. Transport of an organic compound into roots is a function of the lipophilicity of that compound. Lipophilicity can be measured by its relative partitioning between an octanol phase and a water phase. It is expressed as the log of the partitioning coefficient or log K_{ow}. A high log K_{ow} is indicative of low water solubility and high lipophilicity while a small log K_{ow} is indicative of high water solubility and low lipophilicity. There are some differences among roots, but generally, a compound with a higher log K_{ow} value can be adsorbed onto a root in greater quantities [130, 131]. Lipophilicity also implies an equal propensity to partition into soil organic matter. Root absorption becomes difficult with heavily textured soils and soils higher in organic matter.

Compounds that are the most water soluble will be the most available for mass transport and diffusion into roots. However, very soluble compounds may be leached below the root zone since the compounds have no affinity for soil solids. With compounds having lower log K_{ow} values, plant uptake is dependent largely on active plant processes, especially water transpiration rates. Unlike root absorption, translocation to aboveground plant tissues via the transpiration stream (measured as the transpiration stream concentration factor, TSCF) is predominant for compounds having lower log K_{ow}. Plants with higher water transpiration rates can take up more of a compound with a low log K_{ow} than plants with reduced water requirements [7, 22, 132].

Once absorbed into a plant, these compounds may have multiple fates; however, many compounds are bound into plant tissues. Some compounds can be accumulated in root tissue, some can be metabolized in the root, and some can be translocated by phloem and subsequently volatilized from leaf surfaces. Sequestration of organic compounds is a common fate, often associated with cell wall materials such as lignin. Furthermore, plant parameters such as the mean daily water uptake rate have been shown to play an important role in determining the extent of contaminant and water absorption by the root, which, in turn, is related to the contaminant retardation [133]. Volatile compounds or their metabolic products also can be released to the atmosphere through plant transpiration. Many organic contaminants that are recalcitrant in the subsurface environment react and degrade rapidly in the atmosphere with hydroxyl radicals, an oxidant formed in the photochemical cycle.

4 Modeling the Fate and Transport Processes

Mathematical modeling can be used to simplify and clarify the interactions of plants, microbes, and contaminants in soil. Each equation in the model utilizes the concepts discussed in the previous sections in order to predict how plants will affect organic contaminants in soil. These effects can either be a result of interactions with the various physical or chemical fate and transport processes or from biological processes.

4.1 Model Development

For the one-dimensional vertical flow process shown in Fig. 2, the following advective dispersion equation (ADE) can be used to describe the bulk flow of the contaminant [134].

$$\frac{\delta J_c}{\delta z} + \frac{\delta C_T}{\delta t} + r = 0 \tag{2}$$

In Eq. (2), J_c is the contaminant flux ($ML^{-2}T^{-1}$), C_T is the total contaminant concentration (ML^{-3}), and r is the rate of contaminant disappearance ($ML^{-3}T^{-1}$).

The total contaminant concentration in soil can be determined from its components present in the various phases of the soil environment.

$$C_T = \theta_w C_w + \theta_a C_a + \varrho C_s \tag{3}$$

In Eq. (3), θ_w is the volumetric fraction of water, θ_a is the volumetric fraction of air, and ϱ is the soil bulk density. Furthermore, C_w, C_a, and C_s are, respectively, the contaminant concentrations in the soil-water phase, soil-air phase, and soil-solid phase.

In order to solve the ADE, the total contaminant concentration (C_T) should be expressed in terms of the aqueous phase concentration (C_w) by assuming phase

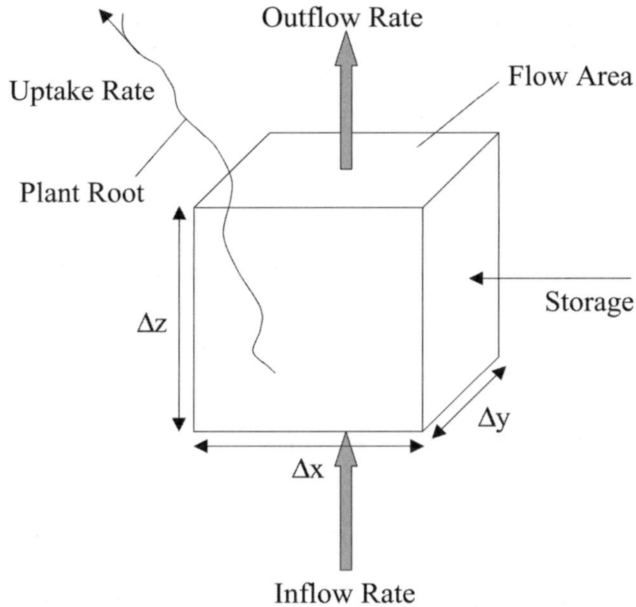

Fig. 2. Elementary volume of soil used in calculating the contaminant mass balance equation

equilibrium. Starting with Henry's law, shown in Eq. (4), the soil-air concentration (C_a) can be directly related to C_w where H represents Henry's constant.

$$H = \frac{C_a}{C_w} \quad \text{or} \quad C_a = H\,C_w \tag{4}$$

Furthermore, according to the soil-water partition law, the sorbed concentration (C_s) can be related to C_w using the soil-water partition coefficient (k_d). This coefficient can be determined from the soil organic carbon mass partition coefficient (k_{oc}) multiplied by the fraction of organic carbon in the soil (f_{oc}). Values of k_{oc} can be found in the literature for a wide range of contaminants [135].

$$k_d = \frac{C_s}{C_w} \quad \text{or} \quad C_s = k_d\,C_w \tag{5}$$

Therefore, Eq. (3) becomes the following, where κ is known as the retardation coefficient.

$$C_T = \theta_w C_w + H\,C_w \theta_a + \varrho\,k_d C_w \tag{6}$$

$$C_T = C_w(\theta_w + H\,\theta_a + \varrho\,k_d) \tag{7}$$

$$C_T = \kappa\,C_w \tag{8}$$

where

$$\kappa = \theta_w + H\,\theta_a + \varrho\,k_d \tag{9}$$

The contaminant flux can be due to advection ($J_{c,adv}$) and diffusion ($J_{c,diff}$). The flux due to advection ($J_{c,adv}$) is the flux of the dissolved contaminant in the aqueous phase due to bulk fluid movement. This is described in Eq. (10) where v_z is the velocity of water moving in the z direction.

$$J_{c,adv} = v_z \theta_w C_w \tag{10}$$

The flux due to diffusion ($J_{c,diff}$) is the flux of the contaminant in the vapor and liquid phases. According to Fick's first law, $J_{c,diff}$ can be determined using the following equation, where D_z is the soil aqueous phase diffusion coefficient and D_{za} is the soil gas phase diffusion coefficient.

$$J_{c,diff} = -\theta_w D_z \frac{\delta C_w}{\delta z} - \theta_a D_{za} \frac{\delta C_a}{\delta z} \tag{11}$$

Therefore, the contaminant flux (J_c) becomes the following after combining Eqs. (4), (10) and (11).

$$J_c = v_z \theta_w C_w - \theta_w D_z \frac{\delta C_w}{\delta z} - \theta_a D_{za} H \frac{\delta C_w}{\delta z} \tag{12}$$

The sink term, r (rate of contaminant disappearance), can be microbial degradation (r_{micro}) and/or uptake of contaminants by plants (r_{plant}). Microbial degradation can take place in both the aqueous and sorbed soil phases. Assuming first order decay constants in these phases, r_{micro} can be expressed as the following summation, where μ_w and μ_s are decay constants for the contaminant in the aqueous and sorbed phases, respectively.

$$r_{micro} = \mu_w \theta_w C_w + \mu_s \varrho C_s \tag{13}$$

Substituting Eq. (5) for C_s allows the equation for the rate of contaminant dis-apppearance to be rewritten as the following in terms of C_w and an overall first order decay coefficient for the contaminant (β).

$$r_{micro} = \beta C_w \tag{14}$$

$$\text{where} \quad \beta = \mu_w \theta_w + \mu_s \varrho k_d \tag{15}$$

Plant uptake can take place only in the aqueous phase. Assuming a first order uptake constant in the aqueous phase, r_{plant} can be expressed as the following, where μ_{wp} is the plant uptake constant for the contaminant in the aqueous phase.

$$r_{plant} = \mu_{wp} \theta_w C_w \tag{16}$$

$$\text{also,} \quad r_{plant} = \gamma C_w \tag{17}$$

$$\text{where} \quad \gamma = \mu_{wp} \theta_w \tag{18}$$

Substituting Eqs. (8), (12), (14) and (18) into Eq. (2) results in the following expression of the ADE equation describing contaminant transport.

$$\frac{\delta}{\delta z}\left[v_z \theta_w C_w - \theta_w D_z \frac{\delta C_w}{\delta z} - \theta_a D_{za} H \frac{\delta C_w}{\delta z}\right] + \kappa \frac{\delta C_w}{\delta t} + \beta C_w + \gamma C_w = 0 \tag{19}$$

Simplifying Eq. (19) shows the form of the ADE equation that can be used to simulate the contaminant transport with advection ($q_z = v_z \theta_w$), aqueous phase diffusion ($D = \theta_w D_z$), gas phase diffusion ($\mathfrak{D} = \theta_a D_{za} H$), retardation ($\kappa$), microbial decay ($\beta$), and plant uptake ($\gamma$).

$$\kappa \frac{\delta C_w}{\delta t} = D \frac{\delta^2 C_w}{\delta z^2} + \mathfrak{D} \frac{\delta^2 C_w}{\delta z^2} = q_z \frac{\delta C_w}{\delta z} - \beta C_w - \gamma C_w \tag{20}$$

Equation (20) quantitatively describes the soil, microbe, and plant interactions that affect the fate and transport of dissolved organic contaminants in soils. The physical or chemical processes included here are retardation, advection, aqueous phase diffusion, and gas phase diffusion. Furthermore, the ADE incorporates biological processes such as microbial degradation and plant uptake.

4.2 Plant Effects on Physical/Chemical Parameters

Just as dissolved contaminant concentrations in the subsurface can affect the presence of plants in the soils due to toxicity, contaminant concentrations in the soil-water phase (C_w) are, in turn, affected by the presence of plants in the soil. Plants can affect contaminant solubility directly or indirectly. As plants transpire, the availability of water in the subsurface to dissolve contaminants decreases. Furthermore, plant root exudates include organic acids that might alter the pH of the soil solution. As soil pH is altered, the functional groups of an organic compound could hydrolyze, resulting in a change in its solubility. Plant roots also harbor microbial colonies that might release biosurfactants that alter the solubility of an organic compound. Each of these interactions between the plant, soil, and microbes affects the value of C_w in the ADE.

The definition of the retardation coefficient (κ, see Eq. (9)), includes parameters for the volumetric water content (θ_w), volumetric air content (θ_a), soil bulk density (ϱ), and the soil-water partition coefficient (k_d). Plants extract water decreasing θ_w which ultimately increases θ_a. Variations in κ along the soil profile directly correspond to variations in θ_w in response to changes in the mean daily water uptake rate by plants [21]. Plant roots also can alter ϱ as described in Sect. 2.1. Plants also add organic matter to soil by not only depositing dead leaves, stems, and roots, but also by exudation from the roots. Increasing the organic matter in soil can alter the partitioning properties leading to a change in k_d. This affects the sorption capacity of the soil for the particular contaminant. Also, plant roots themselves can sorb contaminants. Therefore, plants have an overall effect of retarding the aqueous transport of a contaminant in soil [21].

Plants can also affect advection ($J_{c,\,adv}$, see Eq. 10) by changing the water velocity (v_z) through transpiration. Studies report that transpiration contributes significantly to the transport of organic contaminants in soil [136, 137]. As plants take up water from the subsurface, the horizontal water recharge can lead to an increase in the advective transport ($q_z = v_z \theta_w$) of contaminants by increasing θ_w. However, also due to the plant transpiration, θ_w will decrease, possibly causing the overall advection to decrease. Therefore, the overall rate of advection is dictated by the vertical and horizontal conductivities of soil.

For the transport of contaminant through diffusion ($J_{c,\,diff}$, see Eq. (11)), as plants preferentially extract water, θ_w decreases. However, this results in an increase in the number of soil pores filled with air leading to an increase in θ_a. This can result in a reduction in liquid phase diffusion ($D = \theta_w D_z$) as particular contaminants can equilibrate out of the soil water. Conversely, because θ_a increases, higher gas phase diffusion ($\mathfrak{D} = \theta_a D_{za} H$) can result for a particular contaminant. This gives an opportunity for certain organic compounds to volatilize in the subsurface.

4.3
Plant Effects on Biological Parameters

Plants can transform organic compounds in the subsurface either indirectly by stimulating microbial degradation or directly by taking up contaminants. According to Eq. (15), the decay coefficient (β) is directly dependent on soil parameters such as ϱ, θ_w, and k_d. Furthermore, β is also affected by changes in the first order decay coefficients for contaminants in the aqueous (μ_w) and sorbed (μ_s) phases. Since microbes form aqueous biofilms in soils and plant roots harbor diverse microbial populations with metabolic capabilities to degrade many organic contaminants [4, 10, 17, 20, 24, 124, 125, 128], μ_w and μ_s in the rhizosphere of plant roots are increased.

Several studies show direct plant uptake of several dissolved organic contaminants from soil [132, 138–140]. However, some of the contaminant may be transported back outside of the plant through the roots as exudates or from the stems as part of the sap. The uptake of contaminants (γ, see Eq. (18)) is dependent on θ_w and the uptake coefficient for contaminants in the transpiration water (μ_{wp}). Once inside, the plant has significant metabolic enzymes in both root and shoot tissues. Furthermore, plant metabolic activities can be augmented further by active microbial colonies around roots, in root tissue, in the xylem stream, in shoot and leaf tissue, and on leaf surfaces. These can lead to significant reductions in contaminant concentrations that, in turn, lead to an increase in μ_{wp} along a diffusion gradient. In addition to being metabolized in the tissues, the translocated fractions of contaminants in plants may be stored in plant structures or volatilized from the leaves [7, 141].

5
References

1. Chambers CD, Willis J, Giti-Pour S, Zieleniewski JL, Rickabough JF, Mecca MI, Pasin B, Sims RC, Sorensen DL, Sims JL, McLeam JE, Mahmood R, Dupont RR, Wagner K (1991) In Situ Treatment of Hazardous Waste-Contaminated Soils, 2nd edn. Noyes Data Corporation, Park Ridge, NJ
2. Watts RJ (1997) Hazardous Wastes: Sources, Pathways, and Receptors. John Wiley, New York
3. Anderson TA, White DC, Walton BT (1995) Degradation of Hazardous Organic Compounds by Rhizosphere Microbial Communities, Prog Ind Microbiol 32:205
4. Aprill W, Sims RC (1990) Chemosphere 20:253
5. Cunningham SD, Berti WR (1993) In Vitro Cell Dev Biol (Plant), 29P(4):207
6. Davis LC, Erickson LE, Lee E, Shimp JF, Tracy JC (1993) Environ Prog 12:67

7. Shimp JF, Tracy JC, Davis LC, Lee E, Huang W, Erickson LE, Schnoor JL (1993) Crit Rev Environ Sci Technol 23:41
8. Erickson LE, Banks MK, Davis LC, Schwab AP, Muralidharan N, Reilley K, Tracy JC (1994) Environ Prog 13:226
9. Qui X, Shah SI, Kendall EW, Sorensen DL, Sims RC, Engelke MC (1994) Grass-Enhanced Bioremediation for Clay Soils Contaminated with Polynuclear Aromatic Hydrocarbons. In: Anderson TA, Coates JR (eds) Bioremediation through Rhizosphere Technology, ACS Symposium Series 563, Washington, DC, p. 142
10. Schwab AP, Banks MK (1994) Biologically Mediated Dissipation of Polyaromatic Hydrocarbons in the Root Zone. In: Anderson TA, Coates JR (eds) Bioremediation through Rhizosphere Technology, ACS Symposium Series 563, Washington, DC, p. 132
11. Walton BT, Guthrie EA, Hoylman AM (1994) Toxicant Degradation in the Rhizosphere. In: Anderson TA, Coates JR (eds) Bioremediation through Rhizosphere Technology, ACS Symposium Series 563, Washington, DC, p. 11
12. Cunningham SD, Lee CR (1995) Phytoremediation: Plant-Based Remediation of Contaminated Soils and Sediments. In: Skipper HD, Turco RF (eds) Bioremediation Science and Applications, SSSA Special Publication 43, Madison, WI, p. 145
13. Cunningham SD, Berti WR, Huang JW (1995) Trends in Biotechnol 13(9):393
14. Matso K (1995) Civil Engineering, ASCE, October:46
15. Comis D (1996) J Soil Water Conservation 51(3):184
16. Cunningham SD, Ow DW (1996) Plant Physiol 110(3):715
17. Gunther T, Dornberger U, Fristsche W (1996) Chemosphere 33:203
18. Kruger EL, Anderson TA, Coates JR (1997) Phytoremediation of Soil and Water Contaminants, ACS Symposium Series 664, Washington, DC
19. McIntyre T, Lewis GM (1997) J Soil Contam 6(3):227
20. Wetzel SC, Banks MK, Schwab AP (1997) Rhizosphere Effects on the Degradation of Pyrene and Anthracene in Soil. In: Kruger EL, Anderson TA, Coates JR (eds) Phytoremediation of Soil and Water Contaminants. ACS Symposium Series 664, Washington, DC, p. 254
21. Chang Y-Y, Corapcioglu MY (1998) J Environ Engg, 124(2): 162–169
22. Davis LC, Banks MK, Schwab AP, Narayanan M, Erickson LE, Tracy JC (1998) Plant-Based Bioremediation. In: Sikdar SK, Irvine RL (eds) Bioremediation: Principles and Practice, vol 2. Technomic Publishing Company, Lancaster, PA, p. 83
23. Banks MK, Govindaraju RS, Schwab AP, Kulakow P (2000) Field Demonstration. In: Fiorenza S, Oubre CL, Ward CH (eds) Phytoremediation of Hydrocarbon-Contaminated Soil. Lewis Publishers, Boca Raton, FL, p. 3
24. Anderson TA, Guthrie EA, Walton BT (1993) Environ Sci Technol 27:2630
25. Jury WA, Gardner WR, Gardner WH (1991) Soil Physics, 5th ed, John Wiley and Sons, NY
26. Sposito G (1989) The Chemistry of Soils. Oxford University Press, New York, NY
27. Brady NC (1990) The Nature and Properties of Soils. Macmillan Publishing Company, NY
28. Larson WE, Allmaras RR (1971) Management Factors and Natural Forces as Related to Compaction. In: Barnes KK (ed), Compaction of Agricultural Soils. ASAE Monograph, St. Joseph, MI, p. 367
29. Barley KP (1954) Soil Sci 78:205
30. Bradshaw AD, Chadwick MJ (1980) The Restoration of Land. Blackwell, Oxford, UK
31. Harris RF, Chesters G, Allen ON (1966) Adv Agron 18:107
32. Allison FE (1968) Soil Sci 106:136
33. Jastrow JD (1996) Soil Biol Biochem 28(4/5):665
34. Jastrow JD (1987) Am J Botany 74:1656
35. Tisdall JM, Oades JM (1982) J Soil Sci 33:141
36. Elliot ET (1986) Soil Sci Soc Am J 50:627
37. Gupta VVSR, Germida JJ (1988) Soil Biol Biochem 20:777
38. Cambardella CA, Elliott ET (1993) Soil Sci Soc Am J 57:1071
39. Healy JB, Young LY (1979) Appl Environ Microbiol 38:84
40. Sleat R, Robinson JP (1984) J Appl Bacteriol 57:381

41. Erickson LE, Fan LT (1988) Anaerobic Degradation of Txic and Hazardous Wastes. In: Erickson LE, Fung DY-C (eds) Handbook of Anaerobic Fermentations, Marcel Dekker, NY, p. 695
42. Hutchins SR, Sewell GW, Kovacs DA, Smith GA (1991) Environ Sci Technol 25:68
43. Heider J, Spormann AM, Beller HR, Widdel F (1998) FEMS Microbiol Rev 22:459
44. Karthikeyan R, Bhandari A (2001) J Hazardous Substance Res 3(3):1
45. Vogel TM, McCarty PL (1985) Appl Environ Microbiol 49(5):1080
46. Vogel TM, McCarty PL (1987) Environ Sci Technol 21(12):1208
47. Freedman DL, Gossett JM (1989) Appl Environ Microbiol 55(9):2144
48. Sewell GW, Gibson SA (1991) Environ Sci Technol 25(5):982
49. Conklin AR (1998) Soil Groundwater Cleanup, Feb/March:34
50. Means JC, Wood SG, Banwart WL (1980) Environ Sci Technol 14:1524
51. Banwart WL, Hjasset JJ, Wood SG, Means JC (1982) Soil Sci 133:42
52. Liu H, Amy G (1993) Environ Sci Technol 27:1553
53. Karami-Lotfabad S, Pickard MA, Gray MR (1996) Environ Sci Technol 30:1145
54. Carmichael LM, Pfaender FK (1997) Environ Toxicol Chem 16:666
55. Brock TD (1966) Principles of Microbial Ecology, Prentice-Hall, Englewood Cliffs, NJ
56. Gray TRG, Parkinson D (1968) The Ecology of Soil Bacteria. University of Toronto, Ontario, Canada
57. Wodzinski RS, Coyle JE (1974) Appl Microbiol 27:1081
58. Weissenfels WD, Beyer M, Klein J (1990) Appl Microbiol Biotechnol 32:479
59. Manilal VB, Alexander M (1991) Appl Microbiol Biotechnol 35:401
60. Walter U, Beyer M, Klein J, Rehm HJ (1991) Appl Microbiol Biotechnol 34:671
61. Volkering F, Breure AM, Sterkenburg A, van Andel JG (1992) Appl Microbiol Biotechnol 36:548
62. Westgate S, Bell G, Halling PJ (1995) Biotechnol Letters 17:1013
63. Smith MM, O'Brien TP (1979) Austral J Plant Physiol 6:201
64. Dec J, Bollag J-M (1994) Biotechnol Bioengg 44:1132
65. Schnoor JL, Licht LA, McCutcheon SC, Wolfe NL, Carreira LH (1995) Environ Sci Technol 29(7):318
66. Hissett R, Gray TRG (1976) Microsites and Time Changes in Soil Microbial Ecology. In: Anderson JM, MacFadyen A (eds) The Role of Terrestrial and Aquatic Organisms in Decomposition Processes. Blackwell, Oxford, UK, p. 23
67. Hsu TS, Bartha R (1979) Appl Environ Microbiol 37:36
68. Curl EA, Truelove B (1986) The Rhizosphere. Springer-Verlag, Berlin Heidelberg New York
69. Tracy JC, Marino MA (1989) J Irrig Drainage Engg, Am Soc Civil Engg 115:608
70. Georgen PG, Davis-Carter J, Taylor HM (1991) Soil Sci Soc Am J 55:210
71. Taylor HM, Upchurch DR, McMichael BL (1992) J Plant Nutrition 15(6–7)727
72. Griffin DM (1980) Water Potential as a Selective Factor in the Microbial Ecology of Soils. In: Parr JF, Gardner WR, Elliott LF (eds) Water Potential Relations in Soil Microbiology. SSSA, Madison, WI, p. 141
73. van Gestel M, Merckx R, Vlassak K (1993) Soil Biol Biochem 25(1):109
74. Sommers LE, Gilmour CM, Wildung RE, Beck SM (1980) The Effect of Water Potential on Decomposition Processes in soils. In: Parr JF, Gardner WR, Elliott LF (eds) Water Potential Relations in Soil Microbiology. SSSA, Madison, WI, p. 97
75. Wilson JM, Griffin DM (1974) Soil Biol Biochem 7:199
76. Harris RF (1980) Effect of Water Potential on Microbial Growth and Activity. In: Parr JF, Gardner WR, Elliott LF (eds) Water Potential Relations in Soil Microbiology, SSSA, Madison, WI, p. 23
77. Paul EA, Clark FE (1996) Soil Microbiology and Biochemistry. Academic Press, San Diego, CA
78. Sotomayor D, Rice CW (1996) Soil Sci Soc Am J 60:1822
79. Schmidt EL (1982) Agron 22:253
80. Atlas RM (1981) Microbiol Rev 45:180
81. Alexander M (1985) Environ Sci Technol 18(2):106

82. Erickson LE, McDonald JP, Fan LT, Dhawan S, Tuitemwong P (1991) Annals NY Acad Sci 665:404
83. Atlas RM, Bartha R (1992) Adv Microbial Ecol 12:287
84. Atlas RM, Cerniglia CE (1995) Bioscience 45(5):332
85. Jobson A, Cook FD, Westlake DWS (1972) Appl Microbiol 23:1082
86. Grady CIP, Dang JS, Harvey DM, Jobbagy A, Wang XL (1989) Water Sci Technol 21:957
87. Bouwer EJ (1993) Bioremediation of Chlorinated Solvents Using Alternate Electron Acceptors. In: Norris RD, Hinchee RE, Brown R, McCarty PL, Semprini L, Wilson JT, Kampbell DH, Reinhard M, Bouwer EJ, Borden RC, Vogel TM, Thomas JM, Ward CH (eds) Handbook of Bioremediation. Lewis, Boca Raton, FL, p. 149
88. Greenwood DJ (1975) Ministry Agric Fish Food Technol Bull 29:261
89. Papendick RI, Campbell GS (1981) Theory and Measurement of Water Potential. In: Parr JF, Gardner WR, Elliott LF (eds) Water Potential Relations in Soil Microbiology SSSA, Madison, WI, p. 1
90. Jenkinson DS, Ladd LM (1981) Microbial Biomass in Soil: Movement and Turnover. In: Paul EA, Ladd JM (eds) Soil Biochemistry. Marcel Dekker, NY, p. 415
91. Sparling GP (1985) The Soil Biomass. In: Vaughan D, Malcolm RE (eds) Soil Organic Matter and Biological Activity, Martinus Nijhoff Junk, Dordrecht, p. 223
92. Rittmann BE (1993) Water Resources Res 29(7):2195
93. McCarty PL (1991) J Hazardous Materials 28(1991):1
94. Coleman DC, Reid CPP, Cole C (1983) Adv Ecol Res 13:1
95. Atlas RM, Bartha R (1993) Microbial Ecology: Fundamentals and Applications, 3rd edn. Benjamin/Cummings, Menlo Park, CA
96. Gibson DT, Mahadevan V, Jerin DM, Yagi H, Yeh HJC (1975) Science 189:295
97. Gibson DT, Subramanian V (1984) Microbial Degradation of Aromatic Hydrocarbons. In: Gibson DT (ed), Microbial Degradation of Organic Compounds. Marcel Dekker, NY, p. 181
98. Bossert I, Bartha R (1986) Bull Environ Contam Toxicol 37:490
99. Heitkamp MA, Cerniglia CE (1988) Appl Environ Microbiol 54:1612
100. Heitkamp MA, Franklin W, Cerniglia CE (1988) Appl Environ Microbiol 54:2549
101. Mahaffey WR, Gibson DT, Cerniglia CE (1988) Appl Environ Microbiol 54:2415
102. Heitkamp MA, Cerniglia CE (1989) Appl Environ Microbiol 55:1968
103. Ellis B, Balba MT, Theile P (1990) Environ Technol 11:443
104. Leahy JG, Colwell RR (1990) Microbiol Rev 54:305
105. Morgan P, Watkinson RJ (1990) Water Sci Technol 22:63
106. Grosser RJ, Warshawsky D, Vestal JR (1991) Appl Environ Microbiol 57:3462
107. Lee E, Banks MK (1993) J Environ Sci Health A28:2187
108. Morgan P, Watkinson JR (1993) Biodegradation of Components of Petroleum. In: Ratledge C (ed), Biochemistry of Microbial Degradation. Kluwer, Dordrecht, p. 1
109. Wilson S, Jones K (1993) Environ Pollution 81:229
110. Chainaeau CH, Morel JL, Oudot J (1995) Environ Sci Technol 29:1615
111. Kelley I, Cerniglia CE (1995) J Soil Contam 4(1):77
112. Pradhan SP, Paterek JR, Liu BY, Conrad JR, Srivatsava VJ (1997) Appl Biochem Biotechnol 63–65(0):759
113. Autry AR, Ellis GM (1992) Environ Prog 11:318
114. Reddi LN, Inyang H (1999) Geoenvironmental Engineering: Principles and Applications, Mercel Dekker, Inc, NY
115. Haverkamp R, Vauclin M, Touma PJ, Vacuad G (1972) Soil Sci Soc Am J 41:285
116. Karickhoff SW, Brown DS, Scott TA (1979) Water Res 13:24
117. Piwoni MD, Keeley JW (1990) Basic Concepts of Contaminant Sorption at Hazardous Waste Sites, Ground Water Issue, EPA Document EPA/540/4–90/053, R.S. Kerr Environmental Research Laboratory, Ada, OK
118. Rao PSC (1990) Water Sci Technol 22:1
119. Mendoza CA, McAlary TA (1989) Ground Water 28(2):199
120. Johnson PC, Stanley CC, Kemblowski MW, Byers DL, Colthart JD (1990) Ground Water Monitoring Rev 10(2):159

121. Turco RF, Sadowsky M (1995) The Microflora of Bioremediation. In: Skipper HD, Turco RF (eds) Bioremediation: Science and Applications. SSSA Special Publication 43, Madison, WI, p. 87
122. Mihelcic JR, Luthy RG (1988) Appl Environ Microbiol 54(5):1182
123. Zhang X, Young LY (1997) Appl Environ Microbiol 63(12):4759
124. Sims RC, Overcash MR (1983) Residue Rev 88:1
125. Shann JR, Boyle JJ (1994) Influence of Plant Species on In Situ Rhizosphere Degradation. In: Anderson TA, Coates JR (eds) Bioremediation Through Rhizosphere Technology. ACS Symposium Series 563, Washington, DC, p. 70
126. Watkins JW, Sorensen DL, Sims RC (1994) Volatilization and Mineralization of Naphthalene in Soil-Grass Microcosms. In: Anderson TA, Coates JR (eds) Bioremediation Through Rhizosphere Technology. ACS Symposium Series 563, Washington, DC, p. 123
127. Anderson TA, Coates JR (1995) An Overview of Microbial Degradation in the Rhizosphere and its Implications for Bioremediation. In: Skipper HD, Turco RF (eds) Bioremediation: Science and Applications. SSSA Special Publication 43, Madison, WI, p. 135
128. Reilley KA, Banks MK, Schwab AP (1996) J Environ Qual 25:212
129. Nichols TD, Wolf DC, Beyrouty CA, Reynolds CM (1997) Water Air Soil Pollution 95(1–4):165
130. Briggs GG, Bromilow RH, Evans AA (1982) Pesticide Sci 13(5):495
131. Cunningham SD, Anderson TA, Schwab AP, Hsu FC (1996) Adv Agron 56:55
132. Burken JG, Schnoor JL (1998) Environ Sci Technol 32:3379
133. McFarlane JC (1995) Plant Contamination. In: McFarlane JC, Trapp S (eds) Modeling and Simulation of Organic Chemical Processes. Lewis, Ann Arbor, MI, p. 13
134. Bedient PB, Rifai HS, Newell CJ (1999) Groundwater Contamination, Transport, and Remediation, 2nd edn. Prentice-Hall PTR, Upper Saddle River, NY
135. Dragun, J (1988) The Soil Chemistry of Hazardous Materials. Hazardous Materials Control Research Institute. Silver Spring, MD
136. Narayanan M, Davis LC, Erickson LE (1995) Environ Sci Technol 29(9):2437
137. Zhang Q, Davis LC, Erickson LE (2000) J Hazardous Substance Res 2(4):1
138. Thompson PL, Ramer LA, Schnoor JL (1996) Environ Sci Technol 32:975
139. Burken JG, Schnoor JL (1997) Environ Sci Technol 31:1399
140. Newman LA, Strand SE, Choe N, Duffy J, Ekuan G, Ruszaj M, Shurtleff BB, Wilmoth J, Heilman P, Gordon MP (1997) Environ Sci Technol 31:1062
141. Trapp S, McFarlane JC (1995) Plant contamination: Modeling and Simulation of Organic Chemical Processes. Lewis, Boco Raton, FL

Received: January 2002

The Effect of Plants on the Degradation and Toxicity of Petroleum Contaminants in Soil: A Field Assessment

M. K. Banks [1] · P. Schwab [2] · B. Liu [1] · P. A. Kulakow [3] · J. S. Smith [4] · R. Kim [1]

[1] School of Civil Engineering, Purdue University, West Lafayette, IN 47906, USA.
E-mail: kbanks@ecn.purdue.edu

[2] Department of Agronomy, Purdue University, West Lafayette, IN 47906, USA

[3] Department of Agronomy, Kansas State University, Manhattan, KS 66506, USA

[4] Department of Animal Sciences and Industry, Kansas State University, Manhattan, KS 66506, USA

A field project located at the US Naval Base at Port Hueneme, California was designed to evaluate changes in contaminant concentrations and toxicity during phytoremediation. Vegetated plots were established in petroleum (diesel and heavy oil) contaminated soil and were evaluated over a two-year period. Plant species were chosen based on initial germination studies and included native California grasses. The toxicity of the impacted soil in vegetated and unvegetated plots was evaluated using Microtox, earthworm, and seed germination assays. The reduction of toxicity was affected more by contaminant aging than the establishment of plants. However, total petroleum hydrocarbon concentrations were lower by the end of the study in the vegetated plots when compared to the unvegetated soil. Although phytoremediation is an effective approach for cleaning-up of petroleum contaminated soil, a long-term management plan is required for significant reductions in contaminant concentrations.

Keywords. Phytoremediation, Contamination, Bioavailability, Bioremediation, Petroleum, Toxicity, Microorganisms

Advances in Biochemical Engineering/
Biotechnology, Vol. 78
Managing Editor: T. Scheper

Abbreviations

CFU	Colony Forming Units
EC	Electrical Conductivity
EC_{50}	Effective Concentration for 50% Reduction of Bioluminescence
GC	Gas Chromatography
INT	2-(p-iodophenyl)-3(p-nitrophenyl)-5-phenltetrazolium chloride
LSD	Least Significant Difference
MPN	Most Probable Number
ns	Not Significant
OM	Organic Matter
P	Probability
PCP	Pentachlorophenol
PAH	Polycyclic Aromatic Hydrocarbons
SPT	Solid-Phase Test
TPH	Total Petroleum Hydrocarbons

1 Introduction

Petroleum products are common soil contaminants and often contain potentially toxic compounds, particularly the polycyclic aromatic hydrocarbons (PAHs). Although soil concentrations of petroleum contaminants are highly regulated, these organic compounds may not pose an environmental threat if they are biologically unavailable. Unfortunately, bioavailability is difficult to define unambiguously and equally hard to quantify.

Bioremediation is a common approach for the clean-up of soils contaminated with petroleum products and uses microorganisms to decompose soil pollutants. A variation of bioremediation is phytoremediation, the use of higher plants to further enhance removal the degradation of soil contaminants. Although less costly than other remediation alternatives, bioremediation often does not result in complete removal of contaminants. Furthermore, the fate of the degradation products is often unknown. Bioremediated soils may be deemed as having reached an acceptable end-point if the contaminants can be shown to be biologically unavailable.

1.1 Bioavailability

Bioremediation of hydrocarbon-contaminated soils reduces toxicity, despite the presence of relatively high total concentrations of the contaminants [1]. Regulatory decisions based on contaminant concentration alone may significantly over-estimate the risks [2]. Bioavailability of petroleum contaminants, especially PAHs, in soil is significantly reduced after aging or weathering [3]. Biodegradation can be limited by the sorption of contaminants to soil, incorporation into soil organic material, and by entrapment of contaminants within soil micropores.

1.1.1 Effects of Organic Matter

Growing plants have the potential to decrease the bioavailability of soil contaminants. A large fraction of petroleum contaminants may be strongly adsorbed to organic matter with a very slow rate of desorption. Consequently, the contaminants are not available for biodegradation [4]. Plants increase the organic carbon content of the soil through exudation of soluble organic compounds and sloughing of dead cells and roots. By increasing the soil organic matter content, the plants increase contaminant adsorption and decrease contaminant bioavailability. Increased organic matter content will also leads to increased humification in the rhizosphere. During this process, free radicals in the humus may react with PAHs, forming electron acceptor-donor complexes and initiating covalent binding of PAHs or metabolites to humic material [5].

Increased organic carbon in the rhizosphere also may induce enhanced degradation of petroleum organic compounds, particularly the PAHs. Accelerated dissipation of PAHs was reported to occur after the addition of compost [5]. They noted increased mineralization and formation of non-extractable bound residues and hypothesized that the formation of bound residues represent an important depletion mechanism in soil. Metabolic by-products that are formed from PAH biodegradation (e.g., hydroxy carbonic acids, phenolic or chinoic compounds) can interact chemically with soil organic materials and be incorporated into the humus, thereby forming bound residues that are not bioavailable.

1.1.2
Effects of Soil Physical Characteristics

Physical restrictions also may contribute to reduced bioavailability. Contaminant entrapment may be one of the primary reasons that bioavailability of contaminants is reduced after aging. Non-ionic, hydrophobic contaminants are physically trapped within soil micropores where diffusion is slowed by contaminant partitioning between the solid and liquid phases of the pore. Apparent diffusion rates of contaminant molecules may be low due to the tortuous flow path between the micropore and the outer soil particle.

The diameters of these micropores are less than 1.0 µm and may be as small as 20 nm; consequently, entrapped material would not be accessible to microbes [6, 7]. Roots of all plants have at least some ability to penetrate soil aggregates and small pores. The fine roots of aggressive plants can disrupt soil aggregates, increase exposed surface area, and enhance biodegradation of entrapped hydrophobic contaminants. Root biomass may be as high as 20 g/kg of soil with total root length as high as 100 m/kg of soil [8].

1.2
Phytoremediation

Plants are commonly used to assess toxicity by quantifying plant uptake or observing phytotoxicity symptoms. However, the potential negative or positive effects of plant roots on bioavailability have not been identified. Phytoremediation is showing great promise as a clean-up technology, but environmentally significant pathways of contaminant dissipation have not been completely identified in soil-plant systems. Growing plants also have the potential to decrease the bioavailability of soil contaminants.

1.2.1
Effects of Plants on Contaminated Soils

Plants can interact with hazardous organic compounds through degradation or accumulation [9, 10]. Uptake of the contaminants by the root is a direct function of the pollutant concentration in the soil solution, and involves chemical partitioning on the root surfaces followed by movement across the cortex to the plant's vascular system. The contaminant may be bound or metabolized at any point during transport. Contaminants may be found in plants as freely extractable residues, extractable conjugates bound to plant material, and unextractable residues incorporated in plant tissue [11]. Within a plant, the contaminant may be adsorbed on a cell surface or accumulated in the cell. Many contaminants become bound on the root surface and are not translocated [12].

Plants may indirectly contribute to the dissipation of contaminants in vegetated soil. Soil adjacent to the root contains increased microbial numbers and populations [13]. Furthermore, the number of bacteria in the rhizosphere was found to be as much as 20 times that normally found in non-rhizosphere soil [14]. Short, gram-negative rods (specifically, *Pseudomonas*, *Flavobacterium*, and

Alcaligenes) are most commonly found in the rhizosphere [15]. The increased microbial numbers are primarily due to the presence of plant exudates and sloughed tissue which serve as sources of energy, carbon, nitrogen, or growth factors. The products excreted by plants include amino acids, carboxylic acids, carbohydrates, nucleic acid derivatives, growth factors, enzymes, and other related compounds [16]. The activity of microorganisms in the root zone stimulates root exudation which further stimulates microbial activity [17].

1.2.2
Rhizodegradation

Several published studies have evaluated the effect of plants and the associated rhizosphere on the fate of petroleum contaminants [18–21]. For the most part, the presence of plants enhanced the dissipation of the contaminants. Also, in the studies using ^{14}C-labeled contaminants in closed plant chambers, mineralization is greater in rhizosphere soils than in unvegetated soil, indicating that the bioavailability of the contaminant is increased in the rhizosphere [19, 22, 23].

In a study of the effects of using deep-rooted prairie grasses for the remediation of soil contaminated with PAHs, results suggested that the roots of these perennial grasses may be more effective at stimulating the rhizosphere microflora due to their fibrous nature [21]. Fibrous roots offer more root surface area for microbial colonization than other roots and result in a larger microbial population in the contaminated soil. Big bluestem (*Andropogon geradii*), indian grass (*Sorghastrum nutans*), switchgrass (*Panicum virgatum*), canada wild-rye (*Elymus canadensis*), little bluestem (*Schizachyrium scoparius*), side oats grama (*Bouteloua curtipendula*), western wheatgrass (*Agropyron smithii*), and blue grama (*Bouteloua gracilis*) were evaluated. After 219 days of growth, PAH dissipation was greater in the rhizosphere soil when compared to non-rhizosphere soil. The dissipation ranking among the four PAHs studied correlated with the water solubility of the compound, with the more soluble compound showing the highest degradation.

The mineralization of pentachlorophenol (PCP) and phenanthrene in rhizosphere soil was evaluated in a study using ^{14}C-labeled compounds [19]. Crested wheatgrass (*Agropyron desertorum*) was grown in flow-through plant chambers. In the PCP contaminated soil, 22% of the initial PCP was converted to $^{14}CO_2$ after 155 days in the vegetated soil, while only 5% was mineralized in the unvegetated soil.

The degradation of PAHs in the rhizosphere of a variety of plants grown in petroleum-contaminated soil has been investigated as well [20]. The degradation in the contaminated soils was assessed for different plant types and PAH compounds. Alfalfa (*Medicago sativa*), fescue (*Festuca arundinacea*), big bluestem (*Andropogon gerardii*), and Sudan grass (*Sorghum vulgare sudanense*) were used. Pyrene was one of the target compounds assessed in this study. After four weeks of plant growth, concentrations of pyrene had declined from an initial level of 100 mg/kg to less than 12.6 mg/kg. By 24 weeks, concentrations were less than 2.4 mg/kg in the soil. Pyrene degradation was significantly greater in the presence of plants than without plants (Table 1).

Table 1. Dissipation of pyrene in contaminated rhizosphere soil

Plant species	Pyrene concentration (mg/kg)	
	4 weeks	24 weeks
None	12.6	2.36
Alfalfa	9.9	1.67
Fescue	10.2	1.49
Switch grass	10.8	1.48
Sudan grass	10.7	1.32
l.s.d. ($P < 0.05$)	ns	0.79

ns – not significant.

The fate of benzo(a)pyrene in soil planted with fescue was compared to an unplanted soil [24]. Using ^{14}C-labeled benzo(a)pyrene in sealed plant chambers, more mineralization occurred in the planted soil during the 190 day study. In the planted chamber, 1.3% was mineralized ($^{14}CO_2$), 0.121% was associated with the plant tissue (0.115% with the root fraction and 0.006% with shoots), and 92% of the radiolabel remained in the soil. In the unvegetated chamber, 0.7% was mineralized and 96% of the ^{14}C remained in the soil. Further analysis of the soil revealed that 50% of the original benzo(a)pyrene remained in the unvegetated soil while only 44% of the parent compound was found in the rhizosphere soil.

2 Field Evaluation Protocol

The objective of this research project was to examine the role of higher plants in the degradation of aged petroleum hydrocarbons and reduction in toxicity of contaminated soil. The field site was located in Port Hueneme, California, and used soil that had been contaminated with heavy oil. A field site was constructed, plants established, and the toxicity was assessed over time using germination/emergence tests, microbial toxicity, and invertebrate growth.

2.1 Experimental Setup

A field assessment was conducted using contaminated soil at a U.S. Department of Defense National Test Facility. Soil in the study area was contaminated with diesel fuel and heavier oil contaminants.

2.1.1 *Plot Design*

A contained treatment cell was designed and constructed on-site. Petroleum contaminated soil was transferred to the cell to a depth of approximately 75 cm. The perimeter of the experimental area was outlined by cement rails (typical highway barriers), 90 cm in height. A plastic, geotextile liner was sealed and attached to the cement rails to contain all leachates. The leachates were collected in a sump and pumped into a collection tank. The entire study area was approximately 20 m by 35 m and was divided into 12 rectangular plots in a randomized block design with three treatments (two plant mixtures and an unplanted control) and four replicate blocks.

2.1.2
Soil Characterization

Soil characterization was performed by MDS Harris Agronomic Services in Lincoln, Nebraska (Table 2). The pH and electrical conductivity (EC) were determined on using a saturated extract, inorganic N (NH_4 and NO_3) determined in 2 M KCl extracts, organic matter (OM) by loss on ignition, and texture by the hydrometer method.

The moisture content of each homogenized sample was determined in triplicate by weighing 5 ± 0.2 g of soil into an aluminum dish. After a 48 hour-drying period at 100 °C, the dry matter was weighed and the moisture content calculated.

2.1.3
Plant Selection

Different vegetation regimes selected for the local area were evaluated for their remediation potential by germination studies. Based on germination results, vegetated plots contained one of two plant species mixtures (shown in Table 3). Unvegetated plots were used as controls. Study plots were seeded in amounts typical for each species. Fertilizer was applied during the growing season and irrigation was used as needed.

2.2
Field Analytical Procedures

Soil samples were analyzed for petroleum contaminants, microbial characteristics, petroleum contaminants, and toxicity approximately every three months. Several standard toxicity tests were performed including Microtox (Azur Envi-

Table 2. Selected chemical and physical characteristics of the contaminated soil

pH	7.2
EC	1.2 mohms/cm
NO_3–N	0.4 mg/kg
NH_4–N	1.0 mg/kg
Bray P	6.0 mg/kg
OM	2.1%
Sand	59%
Silt	26%
Clay	15%

Table 3. Plant species mixtures used at the port hueneme field site

Plant mixture #1 (Grasses/Legumes)	Plant mixture #2 (Native grasses)
Bromus hordeaceous (Blando brome)	*Bromus carinatus* (California brome)
Festuca arundinacea (tall fescue)	*Elymus glaucus* (blue wild rye)
Trifolium fragiferum (strawberry clover)	*Festuca rubra* (red fescue)
Trifolium hirtum (rose clover)	*Hordeum californicum* (California barley)
Vulpia microstachys (small fescue)	*Nassella pulchra* (purple needle grass)
	Leymus triticoides (creeping wild rye)

ronmental, Carlsbad, California), earthworm, and seed germination assays. In addition to sampling the soils, plant tissues were analyzed for biomass accumulation, morphological development, and predominance in the plots. All data were statistically analyzed.

2.2.1
Microbiological Quantification

Assessment of microbiological community structure is an important mechanism by which we evaluate bioavailability and soil quality. Comprehensive knowledge of the diversity of the microbial community in ecosystems and their degradative potential is regarded as important when evaluating the endpoint of bioremediation. Microbial communities can be considered as functional units [25], which are characterized by the sum of the metabolic properties of the microbial taxa involved. Consequently, the substrate utilization pattern of the complete community should result from its qualitative (taxonomic diversity) and quantitative (abundance of each taxon) composition and reflect its metabolic potential (utilization of the substrates). Substrate utilization patterns, as recorded with the BIOLOG system, are suitable for rapidly assessing the dynamics of soil communities and evaluating their biodegradative potential.

The Automated Microbial Identification System of BIOLOG, Inc. (Hayward, California) was originally developed for the rapid identification of pure cultures of microorganisms on the basis of their substrate utilization profiles. Direct incubations of environmental samples in BIOLOG microplates with 95 different substrates, produce utilization patterns that are useful in classification and characterization of heterotrophic microbial communities [26, 27]. Also, the community metabolic fingerprint can be used as an indicator of community structure [27]. The use of gram-negative and gram-positive microplates (128 different substrates) allows for the detection of physiological characteristics of microbial communities, and considerable differences can be noted between ecological communities.

In order to obtain substrate utilization patterns of whole rhizosphere and soil communities, microbial suspensions were prepared by extracting rhizosphere and bulk soils. The extract was diluted to a 3 : 10 ratio by saline solution to minimize the possible effects of the contaminants originating from the soil. Each of the test wells of the BIOLOG microtiter plates was inoculated with 150 μl of microbial suspension. The degree to which a particular substrate was utilized can be quantified by measuring the intensity of color change caused by incorporating a Titertik Multiskan Microplate Reader (Flow Laboratories, McLean, Virginia) at 590 nm. The appropriate observation period was determined by the rate at which color develops and the time at which most of the substrates are used. Color development in each well reflects species activity and density as well as the ability of the bacterial community to respond to particular substrates. To directly compare pairs of substrate utilization patterns, simple matching coefficients were calculated [28] and statistical tests performed.

In addition, microbial colonies were enumerated using the Most Probable Number (MPN) Method [29]. Sterile 96-well microtiter plates were prepared for the soil solution. For each plate, three columns served as controls. Ten grams of soil samples are diluted in a saline buffer solution containing 0.1% sodium pyrophosphate (pH 7.5) and 2% NaCl. A 180 μl sterile Bushnell-Hass solution in 2% NaCl was added in each well of 96-well plates except for the control wells. The first twelve rows in the plates were inoculated with 20 μl soil solution (10 : 1 dilution of original soil sample). Ten-fold serial dilutions of up to 10 : 10 were placed in the plates. Then, 5 μl of diesel fuel was added to the wells. The controls were dye, + oil, + and Bushnell-Hass solution; and dye and + Bushnell-Hass solution. The samples were wrapped with Seal-View lab sealing film which allows air transfer but not moisture exchange. The samples were incubated at room temperature for 2 weeks. Fifty μl of filtered sterilized 2-(p-iodophenyl)-3(p-nitrophenyl)-5-phenyltetrazolium chloride (INT) solution of a concentration concentrated at of 3 g/l was added to each well after 2 weeks. The total liquid amount in each well is was 250 μl. In the positive wells, INT is was reduced to an insoluble formazan that deposits intracellularly as a red precipitate. Positive wells were scored after an overnight, room temperature incubation with INT. Enumerations were then based on a pattern of attributed occurrence across a serial dilution [30].

2.2.2
Contaminant Analysis

The following procedure was adapted from previous work [31]. Target contaminant extraction from soil was performed by sieving approximately 3 grams of air-dried soil with a No. 10 (2 mm) sieve, and placing the soil into a tared scintillation vial. A representative moisture content for this sample was recorded. Then, 10 ml of solvent (50 : 50 mixture of methylene chloride:acetone by volume) was added with 100 μl tetracosane in methylene chloride as a matrix spike. The vial was then capped and shaken for 30 minutes. The samples were centrifuged for 10 minutes at 2000 rpm, and the supernatant was poured off into a separate container. This process was repeated two more times, excluding the addition of the tetracosane matrix spike. For every set of samples, at least 10% of the samples were blanks and an additional 15% were replicates. This method was evaluated for extraction efficiency by comparison with soxhlet extraction [31].

Target contaminant concentrations were determined by placing 1.5 ml of the supernatant into a gas chromatography (GC) vial, adding 5 μl of 1000 mg/l androstane in methylene chloride as an internal standard, securely capping, and analyzing this supernatant/androstane mix by gas chromatography (Hewlett-Packard 5890 A gas chromatograph). The chromatograph was interfaced with a Hewlett-Packard Vectra XM2 computer and Chem Station for integration and sampling programming. The GC was equipped with a splitless capillary liner and a flame ionization detector. The DB-TPH column (J & W Scientific, Folsom, California) was 30 m in length with an inside diameter of 0.32 mm and stationary phase film thickness of 0.25 μm. Attached to the column was a guard column constructed of a 2 m length fused silica tubing. A Merlin microseal septum was used to inhibit contamination between samples. Hydrogen was the carrier gas and the source of fuel for the detector and N_2 is was the auxiliary gas. The injection volume for all samples was 2 μl and at an injection temperature of 250 °C. The total run time was approximately 30 minutes. The initial temperature was 40 °C, maintained for 6 minutes after injection. The temperature was taken to 300 °C at the rate of 10 °C per minute to 300 °C and then held for 5 minutes.

2.2.3
Toxicity Assessments

One of the methods used to evaluate the toxicity of the soils was the Microtox Solid-Phase Test (SPT) procedure (Azur Environmental, Carlsbad, California), which consists of 2 two diluent controls and 13 serial dilutions of each sample, all in duplicate. Thirty-five milliliters of the solid-phase diluent were added to a 50 ml-beaker containing 7 g of homogenized sample and a Teflon-covered magnetic bar. This solution was mixed and then incubated in a 15 °C water bath. Using the appropriate pipette and tips, 1.5 ml of the sample suspension were sequentially transferred to the SPT tubes from which serial dilutions were achieved. The content of each SPT serial dilution tube was well mixed before each transfer. The Microtox acute toxicity reagent (reconstituted freeze-dried bioluminescent bacteria *Vibrio fisheri*, 10^6/mL) was added after a 10 minute equilibration time. At the end of 20 minutes, 20 ml of the reagent was transferred to the SPT tubes in the appropriate sequence. The tube's contents of the tube were thoroughly mixed to re-suspend the solids, and column filters were then inserted in all SPT tubes, just above the liquid level. Five hundred microliters of each filtrate were then sequentially transferred to the corresponding cuvettes in the Model 500 Analyzer, and incubated for 10 minutes at 15 °C. The bioluminescence output was read, and the data plotted via software generating log-log graphs of luminescence reduction versus concentration. The results were reported as EC_{50}, which represents the effective concentrations of the soil sample that reduce 50% of the bacterial bioluminescence. The results were adjusted to a dry matter basis.

In addition to the Microtox tests, earthworms, which are considered representative soil macro-invertebrates, were tested. *Eisenia foetida* was used because it is sensitive to many different compounds [32]. A range of contaminated soil concentrations (at least 5 different concentrations by dry weight) was prepared by mixing the soil with an artificial soil composed of 10% sphagnum peat, 20% kaolinite clay, and 70% grade-70 silica sand by dry weight. Soil moisture was adjusted to 35% of the water-holding capacity of the soil. Then, 100 g of soil were placed in plastic ziploc bags that had been punctured 10 times with a thumb tack to allow for the entry of air. Two grams of Magic Worm Food were placed on the surface of the soil in each bag. Adult earthworms were cleaned of soil and hydrated in distilled water and incubated overnight in the water at 20 ± 2 °C for gut purging. A pair of lightly dried earthworms were then weighed and placed in each bag. Each bag was then incubated at 20 ± 2 °C under a light-cycle of 12 hours dark/12 hours light for 14 days. After 14 days, the worms were hydrated and exposed to a repeated gut-purging process. The moist weight of each worm pair was then recorded. Mortality and biomass changes were calculated.

A third toxicity assessment test was performed using seed germination assays. The primary test species used for this type of assay was lettuce, *Lactuca sativa* L. This method estimates acute toxicity of solid material to lettuce in a 120-hour static test. A range of concentrations of test soils was prepared by mixing contaminated soil with an artificial soil (20-mesh, washed silica sand). Soil concentrations were prepared in a dilution series (i.e., 100%, 75%, 50%, 25%, and 0% for the control) by dry weight. Soil moisture was adjusted to 85% water-holding capacity and placed in 16-mm petri dishes. Forty lettuce seeds were then placed evenly throughout each petri dish (at least 0.5 inch from the edge) and covered with 90 g of dry 16-mesh cover sand. The petri dishes were shut and randomly placed in a growth chamber 24 ± 2 °C in the dark for 48 hours. Then, light cycles of 16 hours light/8 hours dark were applied with a light intensity of 4300 ± 430 lux until 120 hours has passed. The number of germinations were then counted, each seed being considered germinated if it sprouted above the cover sand.

2.2.4
Plant Biomass Analyses

Above-ground biomass was measured in two random locations in each vegetated subplot by cutting the biomass in a 0.5 m by 0.5 m frame to within one inch above the ground. The biomass was then dried and weights recorded. In order to quantify

For quantification of root development, soil cores were taken at four random locations in each vegetated subplot. Root samples were taken from site soil using a 15-cm core (300 cm^3 total soil volume). Samples were stored in sealed bags and immediately placed on ice for shipment back to the laboratory. Samples were processed by placing each sample into a 1-L container. Sodium hexametaphosphate was added to approximately half the volume of the container and then filled to volume with hot tap water. Samples were hand-shaken to agitate soil and allowed to soak overnight. The following day, each sample was rinsed well to remove soil particles and suspended in water. Roots and debris were carefully separated by hand. Cleaned root samples were stored in a mixture of 25%/75% percent by volume ethanol/water. Root samples were scanned using the WinRHIZO system (Regent Instruments, Inc., Quebec, Canada) on an Epson 400 scanner. Roots were spread into a single layer on a transparent tray, then scanned and analyzed automatically by the software. Total surface area was determined by the software and standardized to surface area/volume soil. Values for 0 – 30 cm cores were combined to obtain total surface area/volume for each sample.

2.2.5
Statistical Analysis

The data for all tests were subject to analysis of variance (Cohort Software, Berkeley, California) assuming a randomized complete block design with four blocks. Comparisons were made using Fisher's Least Significant Difference (l.s.d.) with a probability less than 0.05 ($P < 0.05$), unless state otherwise. Least significant differences were calculated for all main effects of the treatments and for interactions of the treatments when interactions were found to be significant in the analysis of variance. For the contaminant concentrations, covariate analysis was used to account for differences in the initial contaminant concentrations across the blocks.

3
Results

Experimental observations from the field trial include plant biomass, microbial populations, toxicity, and contaminant concentration data. The field site was evaluated quarterly for all of these parameters.

3.1
Plant Biomass

As shown in Fig. 1, both plant mixtures produced high quantities of above-ground biomass after nine months of growth. The plants in plots with vegetative mixture #1 (grasses and legumes) established faster than those from mixture #2 (native grasses), but that was expected given the growth characteristics of the plant species in each mixture. At the end of the study period, the tall fescue (*Festuca arundinacea*) and rose clover *(Trifolium hirtum)* accounted for more than 85% of the vegetation in the mixture #1 plots. In the plots with mixture #2, red fescue (*Festuca rubra)*, creeping wild rye (*Leymus triticoides)*, and California brome (*Bromus carinatus)* were the predominant species, accounting for more than 90% of the total cover. The dynamics in above-ground biomass reflected the seasons, with strong increases in biomass during growth periods and decreases in biomass during periods of dormancy or slow growth.

Root development followed trends similar to those for plant shoots (Fig. 2; data are not available for months 11 and 14). The plant root structure was fully developed after one year of plant growth in the contaminated soil.

3.2
Microbial Analyses

The three microbial analyses performed clearly illustrated differences between rhizosphere and non-rhizosphere soil.

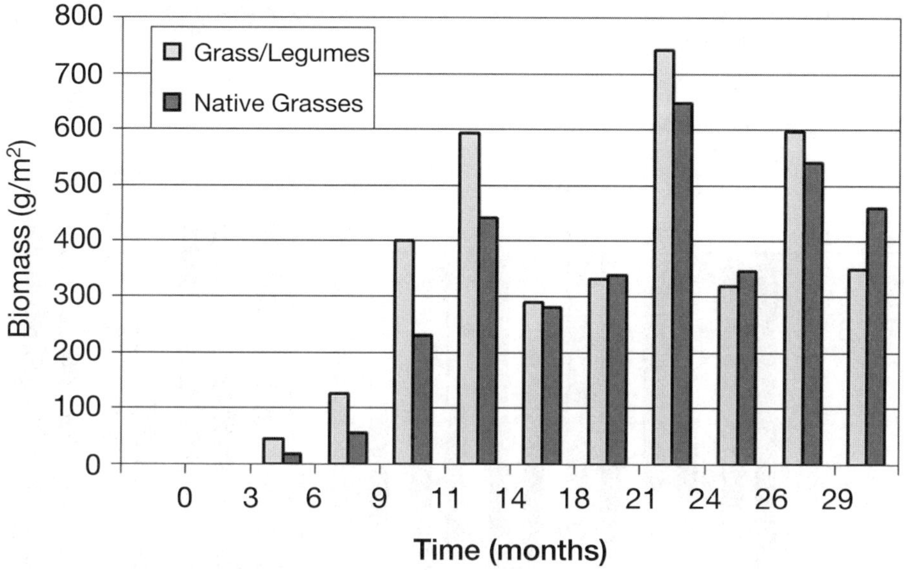

Fig. 1. Above-ground plant biomass at the Port Hueneme Field Trial (l.s.d = 100 g/m^2)

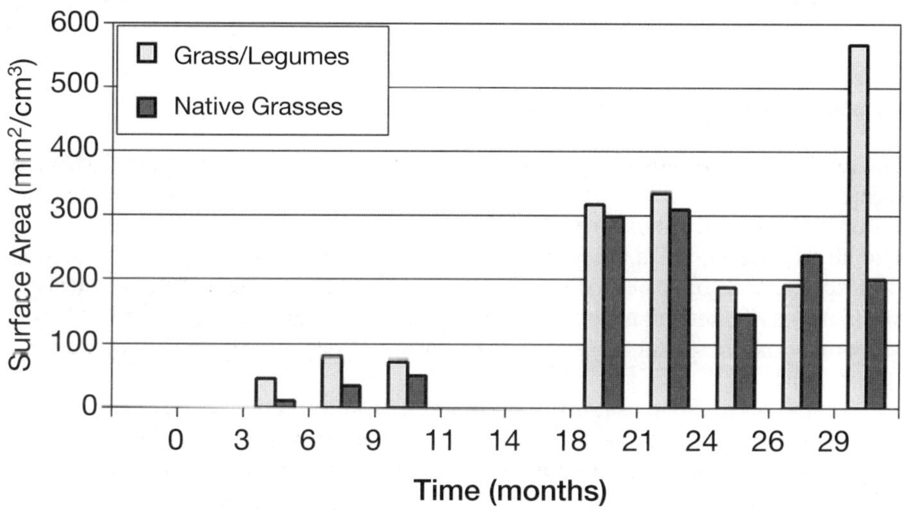

Fig. 2. Root surface area at the Port Hueneme Field Trial (l.s.d = 55 mm^2/cm^3)

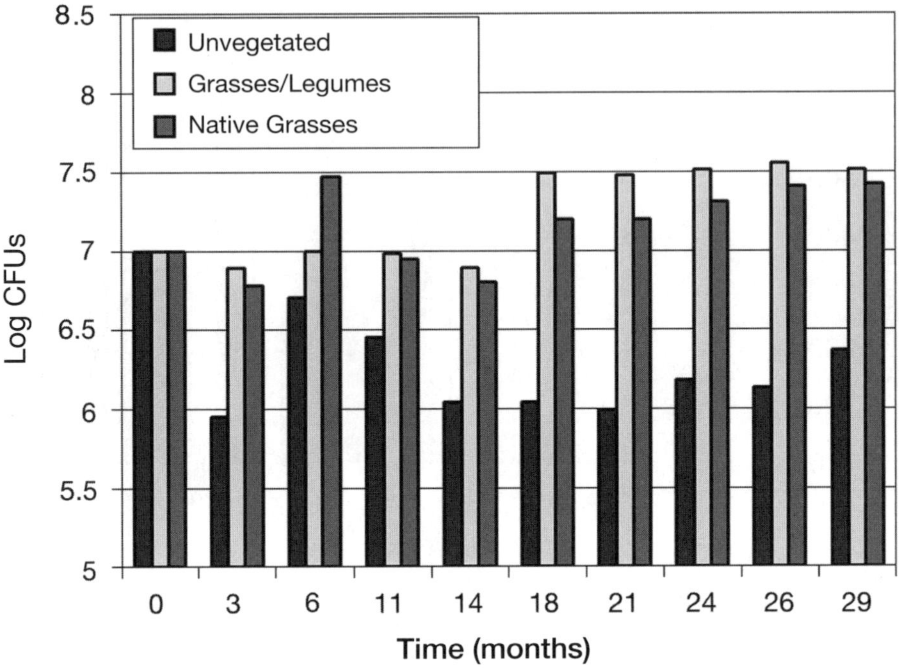

Fig. 3. Total bacterial counts from the Port Hueneme Field Trial (l.s.d = 1.28; log scale)

3.2.1
Microbial Counts

The total bacterial counts (Fig. 3), measured in colony forming units (CFU), were significantly greater in the planted compared to the unplanted soils; however, no statistically significant differences existed between the two vegetated treatments. The large differences in microbial populations between rhizosphere and non-rhizosphere soils have been observed previously [14]. Furthermore, the elevated populations of bacteria in the rhizosphere soils are speculated to be at least partially responsible for increased biodegradation of organic contaminants in vegetated soils.

Targeted organisms, pseudomonads and actinomycetes, also were elevated in the vegetated soil, but the differences were not always significant at $P < 0.05$ (Fig. 4 and 5). Increased *Pseudomonas* spp. populations are typical in rhizosphere soils [15], and these organisms are known to be important in the degradation of petroleum hydrocarbons [33–35].

The populations of actinomycetes, which have been found to degrade PAHs in contaminated soil [36, 37], were not significantly impacted by the presence of plants. The populations were high and apparently unaffected by the rhizosphere.

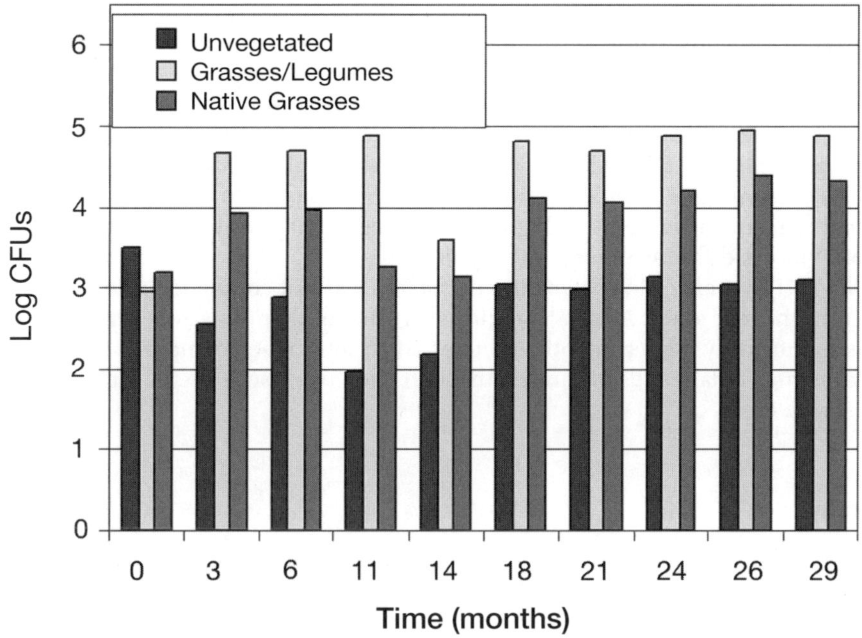

Fig. 4. Pseudomonads from the Port Hueneme Field Trial (l.s.d = 2.1; log scale)

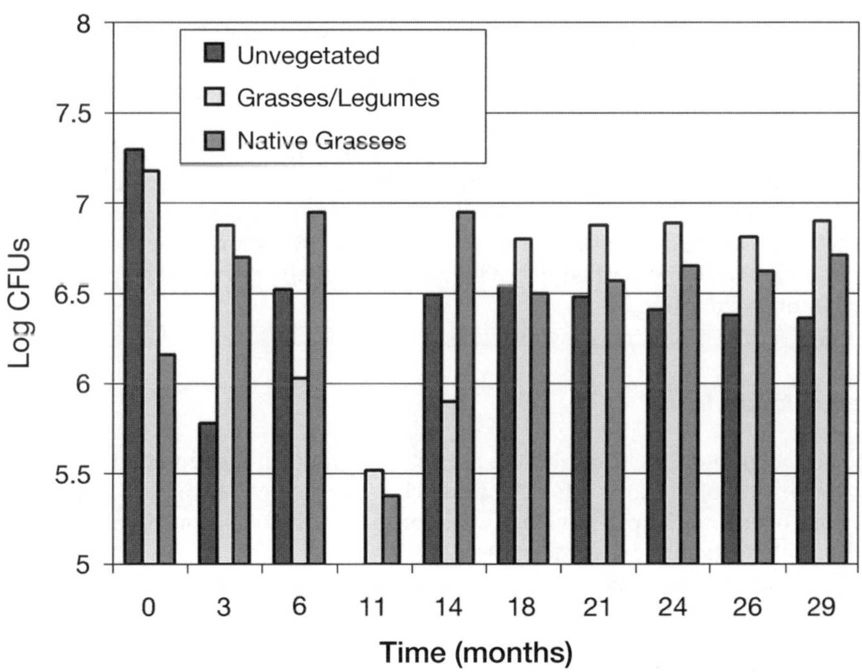

Fig. 5. Actinomycetes from the Port Hueneme Field Trial (l.s.d = 1.02; log scale)

3.2.2
Microbial Community Structure

Microbial community structure was evaluated through the use of BIOLOG and MPN analyses. As shown in Fig. 6, the substrate utilization percentages for microorganisms found in rhizosphere and non-rhizosphere soil were significantly different from three months to nine months of plant growth. However, after nine months, the mean values were higher for the vegetated plots than the unvegetated plots, but the differences were not significant.

The MPN analysis for petroleum degraders was not conducted until eighteen months into the study. At that time, however, the number of petroleum degraders was significantly higher in both vegetated treatments when compared to the bare treatments. This trend continued through the conclusion of the field trial as shown in Fig. 7.

3.3
Toxicity

Various toxicity assessments were performed throughout the two-year study.

3.3.1
Seed Germination Assays

In Fig. 8, germination and emergence of lettuce (*Lactuca sativa*) from undiluted soil samples indicate that the initial toxicity was overcome in all treatments six to nine months after the initiation of the project. It seems unlikely that the initial low rates of germination were the result of residual soil contaminants. Rather, the poor germination was probably due to salt accumulation in the plots. The irrigation water at Port Hueneme was slightly saline, the soil had an elevated salt content from the beginning, and the accumulation of salts near the surface of the soil was visibly apparent. Lettuce is quite sensitive to salt, and germination is impacted even at moderate salt contents of the soil [38]. Management of the salts from the irrigation water coupled with natural rainfall alleviated the problem by the 9-month sampling.

3.3.2
Microtox Toxicity Testing

The impact of both the time and the presence of plants in the remediation system on the toxicity of the soils was determined by measuring the soils EC_{50} values for the soils over time. The EC_{50} values were compared against a control or reference soil sample to evaluate the relative toxicity.

The results in Fig. 9 show that the EC_{50} values slightly decreased for 11 months from the beginning of the study, followed by a significant increase toward the end of the experiment. These two phases are indicative of a slight increase in toxicity (0 – 11 months), and a subsequent reduction (11 – 29 months). However, all of these values are significantly smaller than 2%, which is the hypothetical demar-

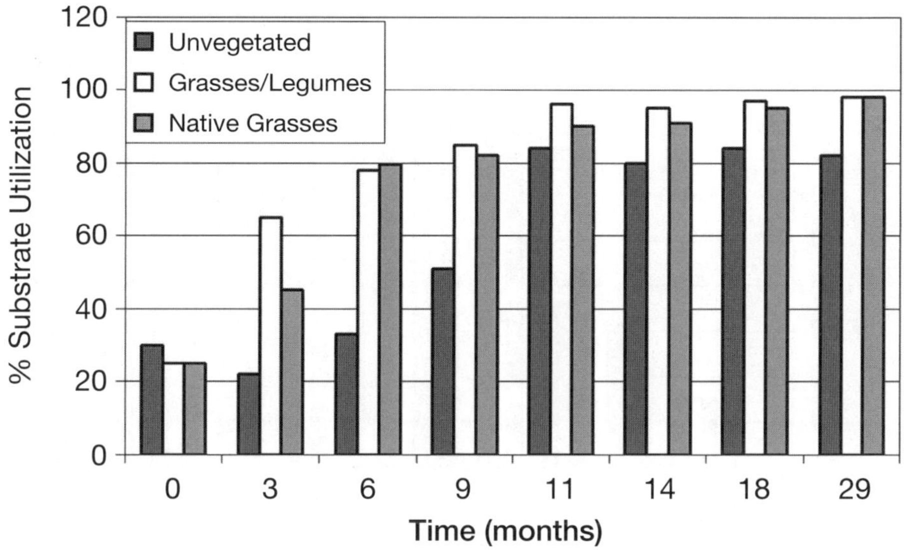

Fig. 6. BIOLOG for the Port Hueneme Field Trial (l.s.d = 14.8%)

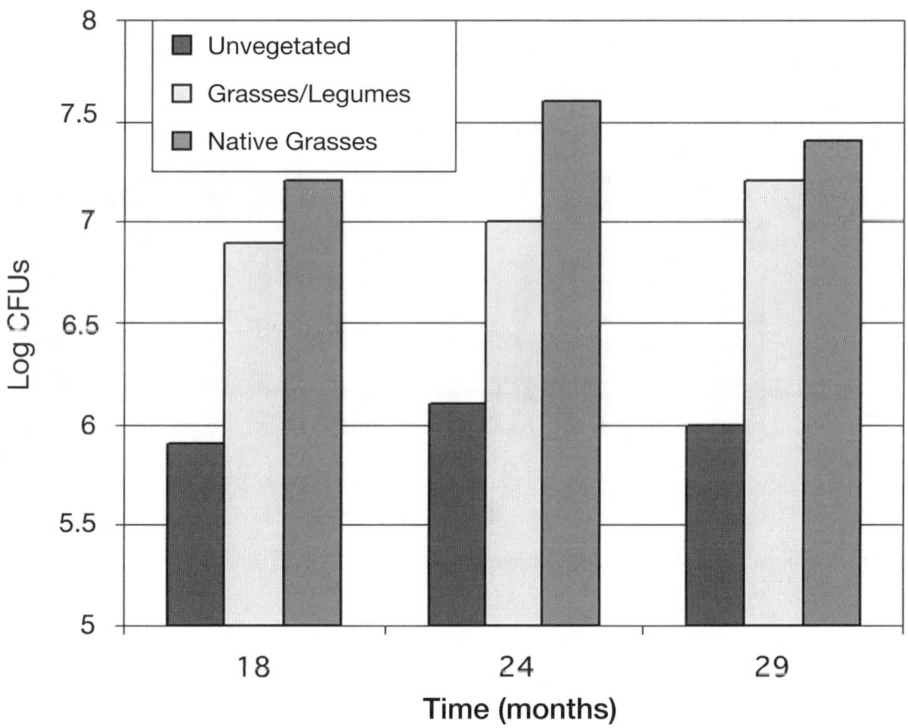

Fig. 7. MPN for petroleum degraders for the Port Hueneme Field Trial (l.s.d = 1.15; log scale)

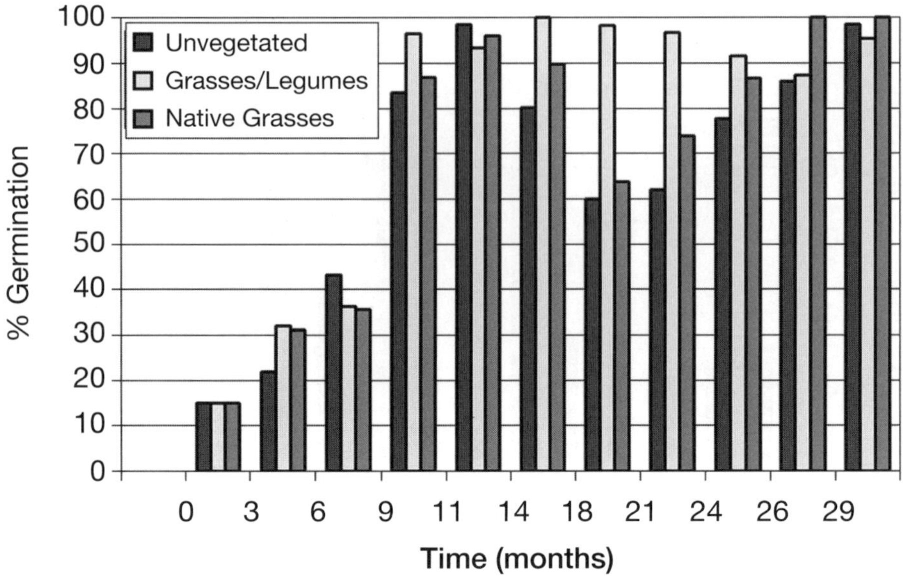

Fig. 8. Germination assessment for the Port Hueneme Field Trial (l.s.d = 32%)

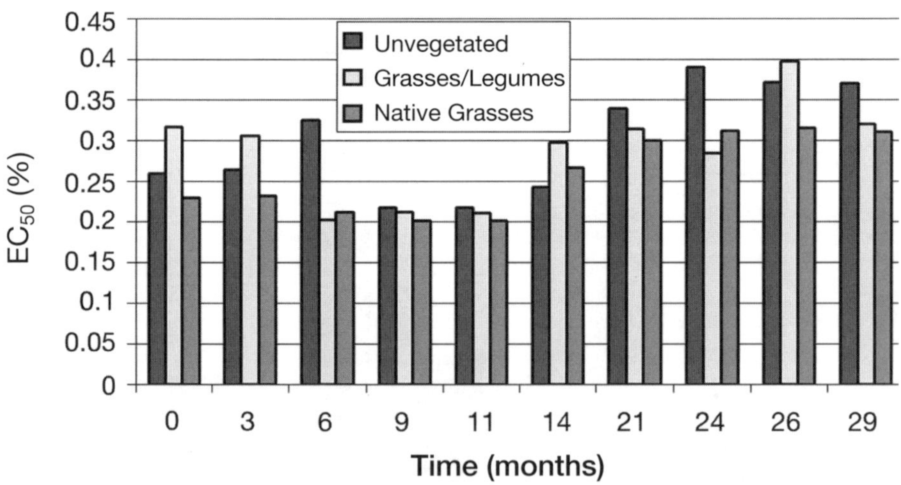

Fig. 9. Microtox analyses for the Port Hueneme Field Trial (l.s.d = 0.1%)

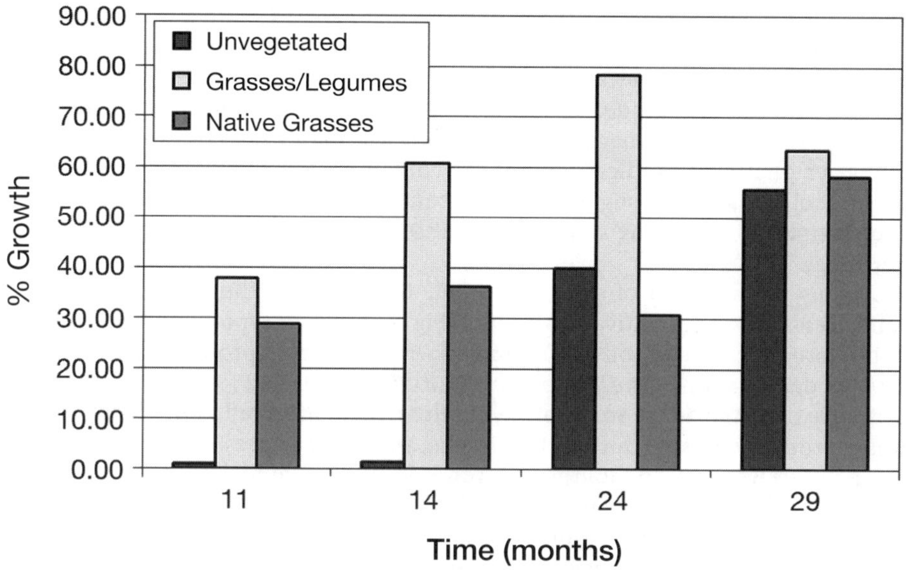

Fig. 10. Earthworm toxicity for the Port Hueneme Field Trial (l.s.d = 38%)

cation between toxic, and non-toxic soils. This would suggest that more time is necessary for the effect of the plants on toxicity to be considerable. In fact, although some of the changes were significant ($P < 0.01$) over time, there was no overall impact of vegetation on soil toxicity as indicated by the EC_{50} values.

3.3.3
Earthworm Assays

As shown in Fig. 10, the earthworm toxicity was not evaluated until eleven months into the study. The undiluted soils were found to be non-lethal to the worms; therefore, we focused was shifted to on the changes in the mass of the worms after the incubation period. The average mass gain by the worms was greater in the vegetated soils at all sampling dates with near zero mass gain in the unvegetated soils at 11 and 14 months. However, the difference is significant only at 14 months. By the end of the field trial, the differences between means were quite small. The presence of plants certainly accelerated the process of improving the suitability of the soils to worm growth, but the passage of time appeared to be an equally important factor as the presence of plants for reducing the toxicity.

3.4
Contaminant Concentration

One of the key parameters being assessed during the field project was total petroleum hydrocarbons (TPH). Preliminary analyses showed that of the total extractable material from a soxhlet extraction, only 10% were non-polar compounds (i.e., petroleum hydrocarbons). The remaining 90% of the compounds were polar materials, typical of a highly weathered material and which are extremely difficult to degrade. Of note is that the leachate from the reactor was analyzed periodically and there was consistently less than 1 mg/l of TPH present in the water phase, also indicating that the contaminants were highly weathered. These preliminary data suggest that the contaminant had undergone extensive transformation, including some humification, prior to the initiation of this experiment.

During the first year of the experiment, dissipation of the TPH followed the same trends in all vegetative treatments (Fig. 11). This is typical for field studies of phytoremediation of soil contaminated with petroleum products [24,39]. Significant degradation of the hydrocarbons occurs simply as a result of the soil disturbance incurred as the soil is placed in the remediation cell [40]. All of the soil was thoroughly aerated during the process, and new surfaces exposed to microbial activity. By fourteen months, the trends for the remainder of the experiment were beginning to emerge. Average TPH concentrations changed very little during the last 15 months in the unvegetated controls, whereas the concentrations continued to decline in the vegetated plots. All differences between unvegetated

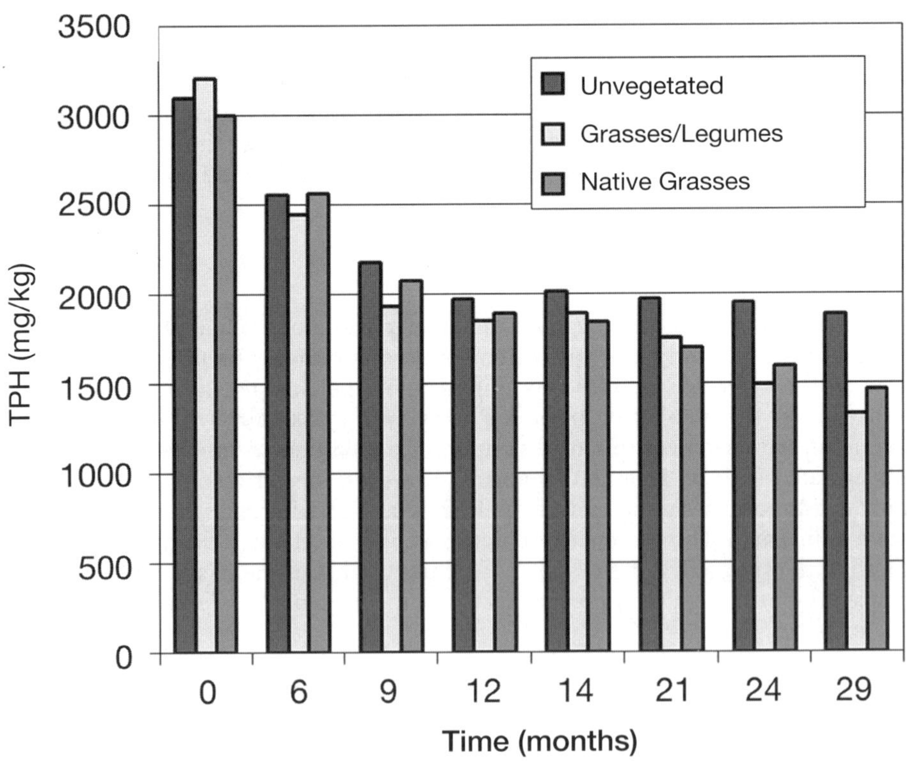

Fig. 11. Total petroleum hydrocarbon concentration at the Port Hueneme Field Trial (l.s.d. = 95 mg/kg)

and vegetated plots were significant ($P < 0.05$) for the last 15 months of the experiment.

4 Conclusions

The plant mixtures chosen for this field trial were not negatively affected by the presence of the contaminants after nine months of growth. Root development was extensive in the second year of the project. Bacterial numbers, petroleum degraders, and plant biomass increased with time, and a strong relationship exists between the presence of plants and the degradation of TPH in our the test plots. These data support the hypothesis that plant roots enhance petroleum degradation by exploring the soil and stimulating microbial activity. As roots grow, they open the soil to better infiltration of water and diffusion of oxygen at depth. Fine roots and root hairs have the capacity to penetrate soil aggregates and micropores, and microorganisms that are attached to the root surfaces are carried with the exploring roots. Thus, the soil microorganisms are able to thrive in regions of the soil that were inaccessible without the roots. Furthermore, their activities are enhanced by higher oxygen concentrations and soluble organic carbon exuded or sloughed by the roots. More petroleum degraders were present in the rhizosphere of the test species than in the unvegetated soil.

The experimental soils were subjected to analysis by toxicity tests, including earthworm growth and mortality, Microtox, and lettuce germination. Some initial toxicity was observed for all tests (particularly the unvegetated soil using the earthworm test), but by the end of the test period no differences in toxicity existed among the treatments. Although plants accelerated the detoxification of the soils, the same results were eventually achieved in the absence of plants. The modest changes in toxicity, particularly in light of the significant reduction in TPH concentrations, reflect the fact that the aged contaminated material was fairly non-toxic from the outset.

Acknowledgement. We would like to thank Karen Miller and Ernie Lory of the US Naval Facility Engineering Service Center for their assistance with site construction, plot maintenance, and experimental design. This research was supported by a grant from the US Environmental Protection Agency.

5 References

1. Loehr RC, Webster MT (1996) J Soil Contam 5:361
2. Parkhurst B, Linder G, McBee K, Bitton G, Dutka B, Hendricks C (1989) Ecological Assessment of Hazardous Waste Sites: A Field and Laboratory Reference. U.S. Environmental Protection Agency Document EPA/600/3–89/013, Corvallis, OR
3. Alexander M (1995) Environ Sci Technol 29:2713
4. Hatzinger PB, Alexander M (1995) Environ Sci Technol 29:537
5. Mahro B, Schaefer G, Kastner M (1994) Bioremediation of Chlorinated and Aromatic Hydrocarbon Compounds. In: Hinchee RE, Leeson A, Semprini L, Ong SK (eds) Lewis Publishers, Boca Raton, FL, p 203

6. Pignatello JJ (1989) Reactions and Movement of Organic Chemicals in Soils. In: Sawhney BL, Brown K (eds), Soil Science Society of America, Madison, WI, p 45
7. Farrell J, Reinhard M (1994) Environ Sci Technol 28:53
8. Detling JK (1979) Perspectives in Grassland Ecology. French N (ed), Springer-Verlag, New York, NY, p 25
9. Finlayson DG, MacCarthy HR (1973) Environmental Pollution by Pesticides. In: Edwards CA (ed), Plenum Press, New York, NY, p 28
10. Schnoor JL, Licht LA, McCutcheon SC, Wolfe NL, Carreira LH (1995) Environ Sci Technol, 29:318 A
11. Bell, RM, Failey RA (1991) Organic Contaminants in the Environment, Jones KC (ed), Elsevier Science, New York, NY, p 189
12. Bell RM (1992) Higher Plant Accumulation of Organic Pollutants from Soils. U.S. Environmental Protection Agency Document EPA/600/SR-92/138, Cincinnati, OH
13. Paul EA, Clark FE (1989) Soil Microbiology and Biochemistry, 1st edn. Academic Press, San Diego, CA
14. Rovira AD, Davey CB (1974) The Plant Root and Its Environment. In: Carson EW (ed) University Press of Virginia, Charlottesville, VA, p 153
15. Barber SA (1984) Soil Nutrient Bioavailability: A Mechanistic Approach, 1st edn. John Wiley, New York, NY
16. Alexander M (1977) Introduction to Soil Microbiology, 2nd edn. John Wiley, New York, NY
17. Barber DA, Martin JK (1976) New Phytol 76:68
18. Reilley K, Banks MK, Schwab AP (1996) J Environ Qual 25:212
19. Ferro AM, Sims RC, Bugsbee B (1994) J Environ Qual 23:272
20. Schwab AP, Banks MK (1994) Bioremediation through Rhizosphere Technology. In: Anderson T, Coats J (eds) American Chemical Society Symposium Series, p 132
21. Aprill W, Sims RC (1990) Chemos 20:253
22. Anderson TA, Walton, BT (1995) Environ Toxicol Chem 14:2041
23. Banks MK, Lee E, Schwab AP (1999) J Environ Qual 28:294
24. Banks MK, Govindaraju RS, Schwab AP, Kulakow P, Finn J (1999) Phytoremediation of Hydrocarbon Contaminated Soils, 1st edn. Lewis, Boca Raton, FL
25. Wuensche LL, Brueggmann L, Babel W (1995) FEMS Microbiol. Ecol 17:295
26. Garland JL, Mills AL (1991) Appl Environ Microbiol 57:2351
27. Garland JL, Mills AL (1994) Beyond the Biomass, Ritz KJ, Dighton J, Giller KE (eds), Wiley-Sayce Publication, London, p 77
28. Sneath PH, Sokal RR (1973) Numerical Taxonomy, 1st ed, W. H. Freeman and Company, San Francisco, CA
29. Wrenn BA, Venosa AD (1996) Can J Microbiol 42:252
30. Klee AJ (1993) J Microbiol Meth 18:91
31. Schwab AP, Su J, Wetzel S, Pekarek S, Banks MK (1999) Environ Sci Technol 33:1940
32. Gibbs MH, Wicker LF, Stewart AJ (1996) Environ Toxicol Chem 15:360
33. Juhasz AL, Britz ML, Stanley GA (1996) Biotechnol Lett 18:577
34. McNally DL, Mihelcic JR, Leuking DR (1998) Environ Sci Tech 32:2633
35. Hoflich G, Gunther T (2000) Bodenkkultur 51:91
36. Kastner M, Breuerjammali M, Mahro B (1994) Appl Microbiol Biotechnol 41:267
37. Madsen EL, Winding A, Malachowsky K, Thomas CT, Ghiorse WC (1992) Microbial Ecol 24:199
38. Bohn HL, McNeal BL, O'Connor GA (1985) Soil Chemistry, 1st edn. Wiley Interscience, New York, NY
39. Nedunuri K, Govindaraju RS, Banks MK, Schwab AP, Chen, Z (2000) J Environ Engg ASCE 126:483
40. Hutchinson SL, Banks MK, Schwab AP (2001) J Environ Qual (in press)

Received: November 2001

Phytoremediation of Heavy Metals from Soils

Terry McIntyre

Environment Canada, 351 St. Joseph Blvd., Hull, Quebec, K1A KKK, Canada.
E-mail: terry.mcintyre@ec.gc.ca

Phytoremediation offers owners and managers of metal-contaminated sites an innovative and cost-effective option to address recalcitrant environmental contaminants. The use of plants or plant products to restore or stabilize contaminated sites, collectively known as phytoremediation, takes advantage of the natural abilities of plants to take up, accumulate, store, or degrade organic and inorganic substances. Although not a new concept, phytoremediation is currently being re-examined as an environmentally friendly, cost-effective means of reducing metal contaminated soil and other substrates throughout North America and Europe. Processes include using plants that tolerate and accumulate metals at high levels (phytoextraction) and using plants that can grow under conditions that are toxic to other plants while preventing, for example, soil erosion (phytostabilization) [1]. Governments worldwide are establishing research and demonstration programs to use this potential [2–6]. Environment Canada has developed a database (PHYTOREM) of 775 plants with capabilities to accumulate or hyperaccumulate one or several of 19 key metallic elements. This chapter addresses key research, potential benefits and limitations, and the potential future needs for phytoremediation. Issues related to intellectual property law, commercialization and public acceptance are touched on.

Keywords. Phytoremediation, Heavy metal accumulation, Metal-contaminated sites, Hyper-accumulation, Bioremediation

Advances in Biochemical Engineering/
Biotechnology, Vol. 78
Managing Editor: T. Scheper

Abbreviations

ATSDR	Agency for Toxic Substances and Disease Registry
MRLs	Minimum Risk Levels
NOAEL	No Observable Adverse Effects Levels
RfDs	Reference Doses
USD	United States Dollar ($)
USEPA	United States Environmental Protection Agency

1 Metals Remediation Market

Contaminated site identification and characterization is an ongoing process. Although there is no comprehensive list of all contaminated sites across North America or Europe, the number of potential sites is staggering. The United States Environmental Protection Agency (USEPA) estimates approximately 500 000 contaminated sites in the United States with over 217 000 in need of immediate remediation [1]. US state-owned or managed contaminated sites are estimated at 19 000 with 38% containing heavy metals and 7% with metals alone [7]. The number of privately owned contaminated sites has been estimated at 24 000 [1]. The costs to clean up the 217 000 most urgent sites in the United States is estimated at about $ 187 billion dollars (USD).

In Canada, based on fragmented and incomplete databases and inventories, the number of contaminated sites has been estimated to be in the range from 20 000 to 30 000, with anticipated costs for clean up of $ 20 billion [8]. In addition, there are approximately 10 000 abandoned mines and 600 abandoned tailing sites. The estimated cost for clean up of abandoned mines alone in Canada has been placed at $ 6 billion. There are approximately 875 million tonnes of mining wastes capable of causing acid mine drainage and approximately 185 million tonnes of radioactive tailings from uranium mines [9].

Globally, the remaining world remediation market value ranged between $ 15 – 18 billion USD in 1998 and it is expected to grow to approximately $ 50 billion USD by the year 2002 [10]. Unfortunately, many questions about definitions preferred methodological approaches and remediation standards for soil contamination remain unanswered in many of the industrialized countries, let alone, those still developing.

1.1 Metals in the Soil Environment

Sources of metals in the environment originate from both natural geological processes and human activities. Natural sources include excessive weathering of mineral and metal ions from rocks, displacement of certain contaminants from groundwater or subsurface layers of soil, atmospheric deposition from volcanic activity, and transport of continental dusts [11]. The most common human routes leading to the introduction of heavy metals into the environment are disposal of industrial effluents, application of sewage sludge, deposition of air-borne industrial wastes, military operations, mining, landfill operations, disposal of industrial solid and liquid waste, use of agricultural chemicals, gas exhausts, and energy and fuel production [12, 13].

The presence of contaminants such as metals in the soil environment, has prompted governments worldwide to initiate environmental laws, policies, and programs to address concerns especially when above ambient natural levels. Specifically, any metal (or metalloid) species may be considered a "contaminant" if it occurs where it is unwanted, or in a form or concentration that causes a detrimental human or environmental effect. Metal concentrations in soil typically range from less than one to as high as 100,000 mg/kg. Irrespective of the origin of the metals in the soil, excessive levels of many metals have begun to pose significant hazards to human, animal, and ecosystem health [14]. This includes the metals/metalloids such as arsenic (As), cadmium (Cd), chromium (Cr), copper (Cu), lead (Pb), mercury (Hg), nickel (Ni), selenium (Se), silver (Ag), and zinc (Zn). These metals/metalloids are those most commonly found at US Superfund sites, contaminated sites identified by the USEPA [15] or comparable Canadian federal or provincial government sites [16]. Other less common metallic species that can be considered contaminants include aluminum (Al), cesium (Cs), cobalt (Co), manganese (Mn), molybdenum (Mo), strontium (Sr), and uranium (U).

Metals in soil can exist either as discrete particles or associated with soil components. This includes the forms listed below [14].

- Free metal ions and soluble metal compounds in the soil solution
- Exchangeable ions sorbed onto inorganic solid phase surfaces
- Non-exchangeable ions and precipitated or insoluble inorganic metal compounds
- Metals complexed by soluble or insoluble organic material
- Metals bound in silicate materials

1.2 Existing Remediation Options

Selection of a remediation option for a site contaminated with metals is complex, time consuming, and site specific. The factors and criteria used in the selection process are listed below [2, 3].

- Size, location, and history of the site
- Site accessibility

- Effectiveness of remediation options
- Soil characteristics (e. g. structure, type, texture, pH, redox potential, cation exchange capacity, etc.)
- Composition, physical state, and chemical state of the contaminants
- Degree of contamination (concentration and distribution)
- Anticipated end use for the site
- Technical and financial resources available
- Regulatory familiarity with technology
- Environmental, legal, and public awareness issues

Traditional remediation options for metals in soil often involve conventional civil engineering practices applied singly, in combination, or in a treatment train. These physical, chemical, and thermal processes are listed below and typically require the physical removal of the metal from the site or the minimization of onsite exposure [2].

- Excavation and retrieval where the contaminated substrate is physically removed and disposed off-site
- Soil washing where the contaminant is removed by size or gravity separation or leaching
- Thermal treatment where the contaminant is volatilized
- Electrokinetics where the contaminants are mobilized as charged particles
- Capping where the contaminated soil is covered with clean soil, clay, asphalt, and/or a geotextile
- Vitrification where the matrix is melted and converted to a glass-like material
- Pneumatic fracturing which utilizes pressurized air injected into the soil to enhance extraction efficiency
- Chemical reduction/oxidation which is the conversion of the contaminant to a more stable or less mobile state

These remediation options are coming under increasing criticism for their high costs, energy intensiveness, site destructiveness, logistical problems, and an increasing degree of public dissatisfaction [2, 17]. These criticisms provide an opportunity for plant-based remediation options with certain plant attributes that are not available with conventional alternatives.

These attributes are listed below [18].

- Potential to transfer moisture from contaminated surface soils and groundwater to the atmosphere
- Production of degradative enzymes
- Ability to survive under a range of ecological conditions
- Solar driven capacities
- Potential to penetrate to significant depths and create large root zones
- Ability to accumulate certain contaminants
- Overall ability to produce and metabolize many natural toxins
- Ability to return the contaminated site to some level of ecosystem health and stability

In addition, other factors have raised the interest in developing alternatives to the high-cost physical, chemical, and thermal techniques for metals-impacted sites.

These factors are listed below and tend to favor the enhanced use of phytoremediation and restoration techniques.

- Sheer volume of contaminated sites containing metals as the principal constituent or as part of mixed wastes [9]
- Ineffective and variable treatment efficiencies under certain conditions associated with traditional physical, chemical, and thermal means for site clean up [17]
- The triple liability that many site owners/managers face including property taxes, clean up liabilities, and possible alternative use value for some sites [19]
- Ongoing demonstrations of plants capable of sequestering, accumulating, and, in some instances, hyperaccumulating up to 19 different metal species [20]
- Volume of metallic waste generated annually by industry [9]
- Human and environmental health concerns associated with a number of existing sites
- Evolving regulatory oversight for specific metals such as As, Hg, and Pb
- Significant potential for reduced costs associated with plant-based remediation techniques [21]

2 Phytoremediation of Metals

Phytoremediation is an emerging technology with considerable promise for remediating and restoring contaminated sites. The continued urgency for contaminated site clean up in developed and developing countries alike, demands that phytoremediation be given careful, serious, and immediate consideration as a cost effective, promising, and innovative environmental technological solution. Phytoremediation is expected, in certain situations, to demonstrate superior economic, technical, and environmental advantages over traditional physical, chemical, and thermal remediation techniques. Given the significant strengths in academic research, government research facilities, and the rapidly evolving capabilities of the environmental biotechnology industry, the North American scientific community is uniquely positioned to draw upon the cadre of silvicultural, agronomic, biological, environmental, and engineering expertise necessary to develop leadership in phytoremediation research. However, the limitations and numerous outstanding issues must be addressed before the utility of this technology can be fully exploited. Some of these issues include scientific and technical research needs, environmental and health risk assessment issues, and the need for regulatory clarification, commercialization, and public acceptance.

2.1 Strategic Approaches for Metals

Phytoremediation, the use of plants or plant products to stabilize or restore contaminated sites, relies on three main strategies for remediating metals including phytoextraction (includes biosorption), rhizofiltration, and phytostabilization [22]. The ability of a plant species to successfully phytoremediate a site contaminated with metals is a function of the amount of metals that can be accumulated by the candidate plant, the growth rate of the plant in question, and the planting density [23].

Phytoextraction relies on plants (both wild and cultivated crop species) that sequester large quantities of toxic elements in the aerial portion of the plant. Hyperaccumulators are able to take up specific metals up to several percent of their dry matter weight [4, 24–26]. Harvested plant parts can be disposed of as small quantities of ash relatively inexpensively. If the toxic elements are of commercial value, they can be recovered through extraction procedures. Plants that have a sizable above ground biomass that can be harvested several times a season to remove toxic elements are the best candidates for phytoextraction. To be most useful for phytoremediation, these plants should also accumulate toxic elements in the order of 2–5% by dry weight within the plant tissues [27].

Biosorption, another form of phytoextraction, refers to the use of biological materials (not necessarily entire living organisms). Several major algal groups (the most common biosorbents) have cell walls with extremely high adsorptive capacities for heavy metals [28–31]. Some flowering plant species can act as very efficient biosorbents, such as the creosote bush (*Larrea tridentata*). When used as a filtering material, this bush has a high capacity for removing Cu [32].

Rhizofiltration is the use of plants with roots that have the capability to absorb and, sometimes, precipitate contaminants from polluted waters. The most effective plants, such as grasses, are those with roots that rapidly regenerate and have large surface areas. Rhizofiltration works best where contaminants are present in low concentrations and in large volumes of effluent. Some plants are able to concentrate toxic metals, on a dry weight basis, to values ranging from 131 to 563 times the concentrations in the contaminant source [33].

Contaminated sites generally support little vegetation. As a result, the barren soils are prone to erosion and leaching which can lead to the off site movement of the pollutants into the environment. Phytostabilization is the use of plants to stabilize the site from the potential migration pathways and depends on the tolerance of the plants to the contaminants. Even if the plants physically remove little or no contaminants, they are useful as phytostabilization agents as long as they tolerate and grow under the contaminated conditions. In many cases, soil amendments are added in order to promote luxuriant plant growth and maximize the stabilization capabilities.

2.2 Advantages of Phytoremediation

Benefits stemming from the use of phytoremediation include in situ application, passive, solar driven 'green' technology, easy to apply, and applicability to a wide range of metals, radionuclides, and organic substances. Furthermore, the broad applicability of the technology allows it to be used at sites not readily remediated by other methods. For example, large sites with low concentrations of contaminants that are widely dispersed at a shallow depth prove to be an interesting challenge for which phytoremediation is particularly suited. The wide knowledge base of agricultural and silvicultural practices can be applied to plant cultivation and management on contaminated sites. Furthermore, several of the prime candidates for phytoremediating metals-impacted sites are well-studied and manageable crop plants. Finally, topsoil can be left in a usable condition with minimum site disruption [34].

Phytoremediation also offers the advantage of eliminating secondary air or water borne wastes that can lead to potential human and ecological exposure pathways. Specifically, toxic substances present in the soil particulate matter may become suspended in the air or water. Plants provide ground cover that stabilizes the contaminated soil and helps reduce the potential spread [35]. By reducing wind blown dust or water erosion, the use of vegetation on the contaminated site helps reduce two important human exposure pathways of direct inhalation from the air and the ingestion of food contaminated as a result of the deposition of the suspended matter onto food plants [36]. Furthermore, exposure of the contaminant to wildlife may also be reduced through phytostabilization. In urban settings, the exposure of children through dermal contact with contaminated soil or from hand-to-mouth exchange of dirt (soil ingestion) could be reduced by the use of phytoremediation as well [37]. Finally, children may also be less inclined to play in areas with substantial plant cover.

Another advantage of phytoremediation is that the plants used may produce recyclable metal-rich plant residues. For example, the tree, *Sebertia acuminata*, could be tapped for the nickel accumulated in its latex. Biomass produced by hyperaccumulators may be disposed of by incineration resulting in an immense reduction of contaminant mass and volume that would need to go to landfills. For example, removing soil contaminated with heavy metals from a two acre site to a depth of 18 inches creates about 5000 tons of soil for disposal. On the other hand, the plant residues produced from phytoremediation would only be between 25 and 30 tons of ash [37]. Recent developments in microbiology demonstrate even greater promise in reducing ash quantities significantly through biological leaching techniques, proven to be effective in reducing a variety of heavy metal constituents [38].

Reclamation of a site using processes such as removal, isolation, incineration to destroy or burn-off contaminants, the return of the residue as sterile material to the site, and the possible supplementation of the soil may cost from $200 to $600 per ton depending on the characteristics of the site and transportation costs. For the removal of one meter of soil from a one acre site, the corresponding costs can be $0.6 to $2.5 million depending on the density of the soil and regional transportation and landfill costs. In comparison, the costs for phytoremediation of the same one acre site including site preparation, planting, and harvesting of plant material may range from only $2000 to $5000 [21]. Furthermore, maintaining a site under cultivation for even ten years is estimated to cost substantially less overall than conventional methods of site clean up. These costs are dependent on site characteristics including whether agricultural equipment and materials are available and can be readily utilized. Such favorable costs could also allow more sites to be cleaned up simultaneously.

Phytoremediation is also seen as a cost-effective means of cleaning up contaminants in groundwater and sediments [21]. For example, the estimated costs for removal of radionuclides from water using sunflowers range from $2 to $6 per thousand gallons including capital and waste disposal costs. Standard microfiltration and precipitation processes cost approximately $80 for the same amount [39].

Phytoremediation offers the additional advantage of making contaminated sites more aesthetically appealing, a bonus that is likely to garner future public interest and support. In fact, a series of public focus group meetings conducted across Canada in 1996 to gauge public perceptions and awareness for environmental applications of biotechnology revealed a high level of public support for the use of plants in a phytoremediation capacity [6].

2.3 Limitations of Phytoremediation

Although phytoremediation offers many advantages over conventional soil remediation techniques, a realistic look at phytoremediation points to some serious limitations. Hyperaccumulating plants often accumulate a specific element only, indicating a limited applicability to sites with multiple of mixed contaminants. Plants have yet to be found for all elements of interest or concern to all countries with soils contaminated by metals. Furthermore, the use of wild or non-indigerous plants for phytoremediation also raises particular issues of risk assessment. Much of the information required for appropriate risk assessment may not be readily available for phytoremediation species. These gaps require more basic herbarium work and field investigations of existing contaminated sites and possibly the establishment of a gene bank to facilitate access to and testing of potential species [40].

Many hyperaccumulator plants are relatively rare, with small populations that often occur in remote areas or are very restricted in distribution [41]. Plants associated with metal-rich soils are often threatened by mining activities. The collection of propagules from hyperaccumulators may also present problems as reproduction of some plants is sporadic and infrequent and seed from native species may be difficult to harvest, separate, and clean. Culturing of these species could be hampered by very specific germination requirements and pollination mechanisms as well as the susceptibility of these species to diseases and herbivores. Most hyperaccumulators have slow growth rates and produce small amounts of biomass [22]. These traits may also be exacerbated by colder northern climates. Some species show considerable variation in shoot production and others may be limited in their efficacy for either strongly or weakly sorbed contaminants. The use of genetic engineering to introduce genes into fast growing cultivars, to regulate root growth, or to increase production of selected plant enzymes may address these problems. Research on this is in progress at the Plant Biotechnology Institute of the National Research Council in Saskatoon and the USEPA Athens Georgia laboratories [42].

Phytoremediation is frequently slower than traditional physical, chemical, or thermal techniques, requiring several growing seasons for site clean up. Therefore, this technology is not an appropriate solution when the target contaminant presents an imminent danger to human health or the environment. Furthermore, the time requirements may prove inhibitory depending on the intended future use of the property. In particular, property developers may not wish to wait the several growing seasons needed for phytoremediation to effectively clean up the site [36]. However, these timing constraints may not be of concern where the site is intended to be restored to its natural vegetated state.

As with any clean up technology, the difficulties associated with waste characterization and the effects of heterogeneous site conditions also apply to phytoremediation [10]. The soil texture, contaminant level, pH, salinity, and toxicity levels must all be within the limits of plant tolerance. Highly soluble contaminants may leach outside the root zone rendering plant uptake less effective. Microclimates on the site may be inhospitable resulting in low numbers of germination and establishment. Biological interactions of species may have a stimulating or a limiting effect on clean up. For example, plants used for clean up may be particularly susceptible to herbivores and pests. Time and effort are needed to develop a complete understanding of biological mechanisms and interactions and to conduct thorough design analyses for biological remediation systems. In general, phytoremediation is most appropriate for treatment of shallow contaminated sites [36] using species acclimated to the local climatological and ecological conditions. However, some species like poplar have extensive root systems that can reach even deep ground water contaminants [43], are applicable to a wide range of contaminants, and are suitable to a wide range of ecological conditions.

This technology may also require that the site be large enough to accommodate traditional agricultural cropping techniques. The terrain of the site and instability of slopes and surface materials will affect efficacy and cost. Phytoremediation may also require considerable input costs such as pretreating waste material or the sites on which the waste is deposited. Artificial chelating agents, irrigation, soil amendments, and insecticides may be required to effectively remediate the site. The techniques for harvesting, processing, and disposing of contaminated biomass must be carefully designed and considered. Further advances in harvesting techniques may yet be required. Sun heating and air drying, ashing or incineration, composting, pressing and compacting, and leaching are all methods that are being considered for disposal [22]. The potential environmental impacts of each approach, however, must be carefully evaluated.

While phytoremediation appears to be a very cost effective approach to clean up contaminated sites, further validation under actual field conditions and cost comparisons are needed in order to allow effective estimates of project costs and process viability on a site specific basis. Study of the economics, performance capabilities, and comparison to competing technologies are still very much required in order to determine the potential share of phytoremediation in the remediation market as well as the potential for commercialization [3, 21, 44].

3 Plant Adaptation to Soils Contaminated with Metal

Botanists have long recognized that some plant species are endemic to soils with high metal and inorganic contamination [5, 41, 45, 46]. Indicator plant species of particular metals have proven useful in prospecting for those metals

[26]. Considerable research has been conducted on the mechanisms of metal tolerance in plants [47] and the establishment of vegetation on contaminated sites [48, 49].

3.1
Physiological Need for Inorganic Elements

As part of their growth and reproductive requirements, plants must not only acquire macronutrients from the soil such nitrogen (N), phosphorus (P), potassium (K), sulfur (S), calcium (Ca), and magnesium (Mg) but also essential micronutrients such as iron (Fe), chlorine (Cl), manganese (Mn), zinc (Zn), nickel (Ni), copper (Cu), and molybdenum (Mo). In order to accomplish this, plants have evolved highly specific mechanisms to take up, translocate, store, and utilize these inorganic nutrients [14]. Explanations for the how plants accumulate and tolerate non-essential metals vary and include inadvertent uptake, the ability to tolerate the metal internally, and/or the subsequent disposal from the plant body (exudation). Several reasons exist why certain plants will accumulate and tolerate these non-essential elements including drought resistance, competition strategy, and pathogen/herbivore defense [50, 51]. Plants can accumulate metals often at levels which are toxic to most living organisms. Therefore, the rationale for plant tolerance and accumulation of metals as a pathogen/herbivore defense mechanism seems to be the most subscribed theory [49, 52].

3.2
Plant Mechanisms for Metal Tolerance and Accumulation

Plants have been likened to solar driven pumps that can extract and concentrate certain elements from their environment [22, 36, 43]. Plant root systems represent an enormous surface area that enables absorption and accumulation of water and essential nutrients for growth. They have metabolic and adsorption capabilities and possess transport systems that can selectively take up many ions from soils [53]. These abilities can be used to remediate metal contaminated sites through a variety of mechanisms. Although the mechanisms by which metal ions enter into the root cells has been the focus of several investigations, it still remains unclear [23].

Most metal tolerant plants exclude toxic metal ions from uptake by exerting a degree of control over the environmental conditions existing in the rhizosphere. This zone contains increased microbial numbers and activity and is supported by plant roots by excreting a variety of compounds that alter the soil-root environment and serve as nutrients and energy sources for the soil microorganisms [42, 53–55]. These microorganisms may also contribute to the remediation of soil contaminants including the symbiotic effects of some fungi on roots [55–58]. Plants and microbes produce and release a variety of active agents and metal chelating molecules (siderophores) that can alter the mobility of metals. Furthermore, plant roots can alter pH by proton extrusion and transfer oxygen to the root zone to facilitate the stabilization or containment of the contaminants by forming stable metal chelates.

On the other hand, some species not only tolerate but actually take up high amounts of toxic metals and other ions. There are a variety of processes for mobilizing metals bound to soil particles including altering the pH, the exudation of phytochemical agents, and specific metal reductases bound to the root plasma membrane [22]. The solar-driven flux created during water uptake can lead to the uptake of these dissolved cations and some anions from the soil, into the roots, and eventually transferred into the plant shoots and leaves. Once inside the plant, the contaminants can subsequently be immobilized in the tissues or, in some cases, volatilized from the leaves. Furthermore, certain metabolic activities from enzymes within the roots, stems, and leafs can detoxify the contaminants by changing the physical state or chemical composition [56].

3.3
Factors Affecting Phytoavailability

The phytoavailability of metals to plants is affected by a number of factors including soil characteristics, environmental factors, and plant considerations [59]. For soil characteristics, the specific factors include pH, hydrous oxides of Fe and Mn, organic matter content, clay content, phosphate content, redox potential, and cation exchange capacity. In terms of environmental factors, climatic conditions, management practices, irrigation, and soil fertilizer practices can all influence the availability of metals to plant uptake. Finally, the species, cultivar, plant tissues, and age of the vegetation can all have considerable effects on metal uptake. Many of these factors can be manipulated through the use of soil amendments or recent developments in plant biotechnology.

3.4
Characteristics of Metal Accumulating Plants

Globally, approximately 775 species of aquatic and terrestrial plants have been identified by Environment Canada in their PHYTOREM database as metal tolerant and accumulating species [20]. Furthermore, the scientific literature contains numerous descriptions of plants that are hyperaccumulators of certain metals. Approximately 400 of these species are known, of which, about 75% are endemic to ultramafic soils and accumulate nickel [45, 60].

Plants identified that could be of potential value in phytoremediation include *Populus deltoides nigra* which takes up atrazine from soils [61], *Alyssum spp.* which accumulate nickel [22], *Aeolanthus sp.* which accumulates Cu and Co [62], *Brassica juncea* which takes up Se [26], and *Alyxia* sp. which accumulates Mn [63]. *Sebertia acuminata,* a species native to serpentine soils in New Caledonia is particularly noteworthy as a hyperaccumulator. The latex of this tree is blue because it contains more than 11% nickel [37].

There are several attributes ascribed to the ideal candidate plant species for phytoremediation of metals. First, the plants should have either a low biomass with a high metal capacity or a high biomass plant with an enhanced metal uptake potential. Specifically, the plant should have a sufficient plant capacity to accumulate the metal of concern within the harvestable biomass at a level greater than 1% (for some metals, greater than 1000 mg/kg). Furthermore, the plant should have a sufficient capacity to tolerate the site conditions and accumulate multiple metal contaminants. Finally, the species should be fast growing and have a suitable plant phenotype for easy harvest, treatment, and disposal [2, 23, 64].

3.5
Plant Biotechnology and Phytoremediation

Plants have evolved a great diversity of genetic adaptations to handle potentially toxic levels of metals in the environment [40]. Biotechnology techniques can be used to create innovative lines of plants and to create new gene combinations to increase the efficacy of phytoremediation capabilities of plants.

The introduction of metal accumulating traits into fast-growing, high-biomass plants would give considerable advantage including the ability to potentially use commonly cultivated plants such as Brussels sprouts and cauliflower [65]. However, the use of plants considered as food crops in this capacity is coming under increasing scrutiny. This approach has been recently shown in the successful transformation of *Arabidopsis thaliana* plants expressing the merA gene, a gene that codes for mercury resistance in bacteria [34].

4
Ecological and Health Risk Assessments

Many candidate phytoremediation species identified in the PHYTOREM database appear to be non-indigenous to North America. Furthermore, the ecological risks associated with the introduction of these alien species into new environments are high profile issues [66]. Specifically, alien species have caused severe ecological and economic impacts by invasion into natural and agricultural ecosystems [67, 68]. However, these ecological risks need to be balanced against the potential health risks stemming from the metal contaminated soil itself. Furthermore, this should include not only the human health risks, but the potential ecological impacts of the contamination on wildlife and the food chain as well. Therefore, a rigorous assessment process investigating the ecological and health impacts of both the phytoremediation system and the contamination should be conducted.

4.1
Ecological Risk Assessment for Introduced Plants

Since remediation sites come in all shapes and sizes, an ecological risk assessment must consider the potential releases of non-indigenous species on small and large scales. The criteria to consider include information on the species to be introduced, the effects of hybridization or cross-breeding, and the potential impacts on the existing species in the receiving or nearby ecosystems. This ecological risk assessment may be of particular importance given the recent attention dealing with the introduction of genetically modified crops. Although the plants used for phytoremediation may not be genetically modified, they are still often species introduced into a new environment. Specifically, phytoremediation may involve the deliberate sowing of an alien species beyond its current range coupled with the simultaneous manipulation of the environment. Finally, consideration of the information needed to conduct a thorough ecological risk assessment and how it can be reasonably obtained needs to be addressed.

For the introduced species, information will be needed on the plant taxonomic identification, reproductive biology, sexual compatibility with related species, life history characteristics, and ecological niche. If the plant to be introduced contains a deliberately introduced gene, then details of the donor organisms and methods of incorporating the gene into the host plant will be required as well. Therefore, a starting point for any risk assessment when introducing a plant into a new environment is an understanding of plant systematics. For some species, taxonomic identification may be simple while other species, such as crop plants, weeds, and ornamentals, may exhibit complex hybridization, reproductive anomalies, and odd distribution patterns that would make them difficult to analyze. Furthermore, the classification nomenclature of some species may be inadequate and confusing [69]. For many plant species, the risk assessment may be hampered by fundamental gaps in the necessary information. However, data on their basic reproductive biology, life history characteristics, and ecological niche may be generated through controlled laboratory and field research.

The ability of an introduced plant to interbreed with weedy and wild relatives is of concern, particular for genetically modified plants. Cross pollination must be considered in terms of the sexuality of the plant, the presence of compatible relatives, the breeding systems of the plants, flowering phenologies, common means of pollination and fertilization, and the ability to set viable seed under field conditions [70]. Furthermore, the invasiveness of the non-indigenous plants through seed or other propagule dissemination through wind, water, insect, or animal means needs to be considered. However, obtaining this information for phytoremediation species that are not typically used in agricultural or forestry could be formidable.

The ecosystem into which a plant is released may range from a microhabitat of a few square meters to an ecozone that ranges over thousands of hectares. It will be important to adopt a uniform scale of classification that is relevant to the risk assessment process for plants used in phytoremediation. The characterization of the receiving environment exposed or potentially exposed to the plant presents another considerable challenge. This requires knowledge of the biotic and abiotic features within the ecosystems and the existing interrelationships between the various features. This would even require an understanding of the potential ecological impacts at the genetic and population levels as well in order to determine the potential of the plant to interbreed with related species (hybridize). Furthermore, once the plant or its hybrids become established and spread in the ecosystem or nearby ecosystems, the potential to cause ecological degradation by disrupting the existing ecosystem processes will need to be addressed.

Rare and endangered species in the current ecosystem must also be considered. Status reports and databases for rare species are being developed in Canada [71]. However, it may prove difficult to obtain up-to-date and accurate information on some rare species for risk assessment purposes. Furthermore, it may not be possible to easily obtain information on the biology of those species to determine how they may interact with or be affected by the plant introduced during the phytoremediation of a contaminated site.

Phytoremediation raises a related issue if the source of starting material for phytoremediation is itself a rare plant. Botanists who study rare species do not encourage the collection of specimens from its natural habitat unless the population can withstand the loss. An unacceptable loss may be set at more than 4% of the individuals in the population [71]. When collecting rare species for use in a phytoremediation system, practitioners will have to exercise caution to insure that they are not destroying the population in its natural habitat.

Although there is no universal approach to predicting invasiveness, several approaches used in combination will increase the accuracy of predictions [72]. Further work is needed to determine the mechanisms whereby plant community diversity within an ecosystem may influence the potential for invasion [73]. A further challenge for researchers in this area will be to integrate the research at genetic, population, and ecosystem levels [68]. In any case, the risk assessment procedures needed to predict the potential of a plant to establish in an ecosystem, invade new ranges, persist, and spread include the following [72].

- Groupings or lists of species that have become naturalized beyond their range
- Traits of weeds and ruderal species to predict invasiveness
- Predictions of invasiveness based on home range climates
- Models that predict a species rate of spread in a new range
- Results from controlled growth facilities and greenhouse experiments
- Comparison of performance of congeners especially through field trials

4.2 Health Risks Associated with Metal Toxicity

Although several mechanisms may be involved and the process for each metal different, the general toxicity of metals occurs by interfering with biochemical or homeostatic processes in the cell. Metal ions may also be involved with several chemical reactions in the cell that produce highly reactive oxyradicals that can interfere with cellular biochemistry or bind to genetic material. The effects associated with each metal, however, depend on the site in the body in which it accumulates (i. e. the target organ), the particular chemical pathway that it disrupts, the chemical form of the metal, and, sometimes, its oxidation state. The ultimate result is the death of cells. Although there is a constant turnover of cells within any tissue, the toxicity of a metal is only expressed if a large enough number of cells are affected to cause a loss of function of the tissue or organ.

Some metals, such as Cu and Zn, are important co-factors for enzymes that are essential for normal metabolism. These metals have normal regulatory enzymes controlling their levels in the cell where the concentrations tend to remain stable. Generally, essential metals are only toxic at relatively high concentrations. Other metals, such as Hg, Cd, and Pb, are not essential and have no specific reg-

Table 1. Summary of the Effects of Metal Exposure in Mammals [79, 80]

Metal	Primary toxic form	Effects
Cd	Cd^{2+}	kidney dysfunction, osteoporosis, hypertension, reproductive effects
Cr	Cr^{6+}	carcinogenicity, teratogenicity, chromosomal aberrations, liver and kidney damage
Cu	Cu^{2+}	liver disease, brain damage, bone disease
Hg	methyl-Hg	neurological, reproductive effects, kidney (as Hg^{2+})
Ni	Ni^{2+}	brain, liver, kidney effects, carcinogenic, reproductive effects
Pb	Pb^{2+}	neurological, hematological, immune, kidney effects
U	UO_2^{2+}	kidney disease, minor liver and muscle effects
Zn	Zn^{2+}	growth retardation, anemia, reproductive effects

ulatory mechanism except for the general sequestering enzymes like metallothionein. These metals tend to be concentrated to a greater extent and may become toxic at lower tissue concentrations.

A wide range of effects has been attributed to specific metals, some of which are shown in Table 1. Kidney disease is a common effect of exposure to As, Cd, Pb, Hg, and U [74–77]. The function of the kidney is to help control and regulate the homeostasis of ions and fluid in the blood [78]. Kidneys that are affected by metals lose their ability to re-absorb ions and other materials before they are lost in urine. This can result in additional effects to the organism such as osteoporosis which results from the loss of bone mass. The most common method to assess the effects after metal exposure and to determine if the kidney is performing adequately is to measure the levels of small molecular weight proteins, glucose, and other ions in the urine.

Another major effect attributed to metals is on genetic material, manifested as mutagenicity, teratogenicity, or carcinogenicity [81]. Hexavalent chromium (Cr^{6+}) and Ni are considered carcinogenic [82], probably because of the production of oxyradicals that interfere with general cell functioning [83, 84]. On a large scale, high levels of oxidants can lead to cardiovascular disease, immune disease, and cancer [85]. Several metals are known to be toxic to embryonic and fetal tissues and may be teratogenic [86]. Damage to chromosomes has been observed in humans exposed to Pb and Cd [87].

The ecological significance of genetic damage in wild populations is unclear. Recent reviews of environmental genotoxicity [88, 89] indicate that there can be a loss of population fitness with large changes in the genome caused by chemicals (or other factors). These changes to the genome could reduce the ability of the affected population to respond to environmental and ecological stresses. Although studies have documented genetic effects from metals and other contaminants in a number of species [90], the significance of the effects in terms of the health of the population has only recently been the focus of research.

4.3 Impacts on Wildlife and Accumulation in the Food Chain

One of the greatest concerns regarding the phytoremediation of metal-contaminated soils centers around the potential impact on wildlife and the accumulation in the food chain. Research has clearly established the linkage between the levels of contaminants in the soil and plants and the concentrations of those compounds in the tissues of wildlife feeding in the area. Several examples of elevated metal levels in animal tissues as a result of exposure to a contaminant source are summarized in Table 2.

The toxicities of most metals to mammals and birds are well known. A large amount of ecological toxicology data is available from the Environmental Sciences group at Oak Ridge Laboratories who implement the recommendations by the USEPA into field applications [96, 97]. These field studies have documented reduced kidney function (Cd, Pb), dental lesions (F), chromosomal aberrations (Cr), neurological signs (Hg), and increased relative tissue weights (Pb) in species living in metal-contaminated environments. Secondary effects include the poisoning of predators feeding on highly exposed mammals and birds, interactions between metals that produce deficiencies in essential nutrients, and secondary diseases caused by the dysfunction of the target organ (e.g. malnutrition caused by the onset of dental lesions).

Of particular concern is the potential impact of metals and food chain accumulations in herbivores and pollinator species. In Canada, one of the best known examples is the accumulation of Cd in deer and moose [98]. This is probably caused by the ingestion of plants growing on Cd-mineralized rock [99]. Whether by natural processes or under the influence of a point source, a common feature is the presence of high burdens of metals in populations living in environments containing high concentrations of the contamination. This strong correlation has led to the recommendation that mammals be used not only as monitors of the environmental conditions [92, 100, 101], but as sentinels for environments where humans may be affected by the high contaminant levels [102].

These detailed studies have also led to recommendations that wildlife species should not be subjected to exposures that result in tissue levels above specific concentrations because of the likelihood of effects (e.g. kidney concentrations of greater than 30 mg/kg for Cd or Pb). Methods are also available to extrapolate No Observable Adverse Effects Levels (NOAEL) from laboratory studies in order to estimate safe exposure levels in wild populations. Furthermore, handbooks and guidelines [79, 80, 103] and online databases, such as MEDLINE and TOXLINE, give information on metal toxicology. Taking this a step further, the Agency for Toxic Substances and Disease Registry (ATSDR) also provides profiles that summarize data on the toxicity of metals to humans, based on animal tests. They also provide minimum risk levels (MRLs), similar to reference doses (RfDs) for health guidance on specific metals.

The ability to predict the exposure of biota at contaminated sites has shown considerable advancement in recent years with the development of ecological risk assessment methods. The exposure of a species can be estimated by calculating the amount of food, water, and soil ingested daily and then comparing to an RfD

Table 2. Examples of High Tissue Burdens of Metals in Contaminated Environments

Element	Species	Location	Contaminant source	Tissue	Mean (mg/kg wet weight)	Reference
Cd	whitetail deer	Pennsylvania	zinc smelter	liver	2.9 vs 0.5 in control	91
Cd	whitetail deer	Pennsylvania	zinc smelter	kidney	23.5 vs 3.9 in control	91
Pb	whitetail deer	Pennsylvania	zinc smelter	liver	0.2 vs 0.05 in control	91
Pb	whitetail deer	Pennsylvania	zinc smelter	kidney	0.5 vs 0.2 in control	91
Hg	wood mouse	England	chloralkali plant	liver	0.23 vs 0.04 in control	cited in 92
Ni	moose	Finland	industrial activity	kidney	0.19 (site 1)/0.12 (site 2) vs 0.03 in control	93
Cu	vole	Ontario Canada	Cu-Ni/U tailings	liver	7 (Cu tailings)/4 (U tailings) vs 3.5 in control	94
Cu	short-tailed vole	England	refinery	whole body	Concentrations near refinery 50% greater vs control	95
Cd	short-tailed vole	England	refinery	whole body	Concentrations near refinery 50% greater vs control	95

value. More advanced risk estimation procedures use spatial (e.g. home range) and temporal variability (e.g. seasonality) to improve the accuracy of the estimates.

Large-scale, multi-disciplinary risk assessment studies of biotic communities living on contaminated sites have reported that small mammals and birds feeding on soil invertebrates tend to be the most exposed species. For phytoremediation sites, the additional pathway of plant-to-herbivore or plant-to-omnivore is also a major exposure pathway. While it has been suggested that the metal content in hyperaccumulators would render these plants unpalatable to grazers, this issue requires much more consideration especially in regard to the accumulation of toxins in the food chain [3, 36]. Understanding these pathways more fully, prior to the construction of the sites, would probably help in the management of potential effects in exposed species.

The potential impacts of remediation on wildlife is a continually evolving field. Prediction of impacts for any particular site must include information on the physical and chemical characteristics of a site, the wildlife species, and the contaminants involved. Canadian exploration of phytoremediation potential indicates contamination of some wildlife groups is likely. However the effects can be monitored and managed through environmental monitoring programs including the volatilization of metals or other contaminants by plants. Modeling the exposure and impact prior to constructing the site and monitoring of the physical, chemical, and biological environments after construction is completed, can aid in developing management procedures that allow the implementation of phytoremediation while minimizing the effects on resident wildlife.

By conducting these prior assessments, the design of remediation sites can reduce potential exposures to wildlife. Physical barriers such as fences can be used to stop large animals but not ground hogs, muskrats, and other burrowing animals. Nor will these obstructions prevent birds from entering into the contaminated site. To stop these other vectors, methods such as netting, electrical, or sonic devices may need to be employed. The use of steep banks on small ponds will help reduce the littoral zone and hence reduce the presence of aquatic invertebrates and small fish. This will in turn, reduce the presence of wildlife attracted to these species. Because of the significance of the soil invertebrate transport pathway, it may be necessary to harvest hyperaccumulating plants in order to reduce the production of metal-rich litter and organic material that accumulates on the soil. These deterrents and landscape features must be incorporated into the early stages of any remediation plan since re-engineering the site would be very expensive.

5 Status of Phytoremediation Applicability to Metals

Several phytoremediation techniques are commercially available or currently under development. Results from demonstration projects in Canada, United States, and Europe, coupled with dialogue from representative scientists actively engaged in the field, have identified the commercial readiness of a number of phytoremediation techniques for specific metals of concern [3, 4, 20]. These are summarized by inorganic constituent in Table 3.

Table 3. Relative Commercial Readiness of Phytoremediation Techniques for Specific Metal Contamination (Ranking: 0 = none; 1 = basic research underway; 2 = laboratory scale/ready for field; 3 = field scale deployment; 4 = under commercialization)

Metal Contaminant	Ni	Co	Se	Pb	Hg	Cd	Zn	As	Cs	Sr	H^3	U
Readiness	4	4	4	4	3	2	3	1	3	2	3	3
Uninduced Phytoremediation	4	4	4	0	4	4	4	1	2	2	n/a	0
Induced Phytoextraction	0	0	0	4	0	0	0	0	0	0	0	3
Regulatory Acceptance[1]	Y	Y	N	Y	N	Y	Y	N	Y	Y	Y	N

[1] Regulatory acceptance of technique under development.

5.1 PHYTOREM Database

The Environment Biotechnology Applications Division of Environment Canada in Hull, Quebec has compiled a database of worldwide terrestrial and aquatic plants that have potential value for phytoremediating sites contaminated with metals. The database was built using an extensive computerized search of public accessible databases and commercial abstracting services involving keyword searches cross-referenced with 19 elements for which data was required. The keywords and phrases used in the searches are listed below while the specific elements are listed in Table 4.

> *Keywords:* Biomarkers of Toxic Exposures, Bioremediation, Biotransformation, Constructed Wetlands, Decontamination, Heavy Metals, Heavy Metal Accumulation, Heavy Metal Tolerance, Hyperaccumulate, Microbial Bioremediation, Phreatophytes, Phytoextraction, Phytoremediation, Phytostabilization, Rhizofiltration, Rhizosphere Biodegradation, Serpentine, Sorption of Heavy Metals, Toxic Metals, Transgenic Plants and Phytoremediation, Treemediation, Trees and Phytoremediation, Vegetative Remediation.

The preliminary database resulting from these searches contains 775 species records covering 76 families, 39 orders, and 9 subclasses. This includes not only vascular plants further categorized into flowering plants, conifers, ferns, and their allies, but other species such as bacteria, algae, lichens, fungi, and bryophytes (mosses and liverworts) that are particularly useful as complementary sorbents.

Table 4. List of Elements Included in the PHYTOREM Database

Aluminum (Al)	Cobalt (Co)	Mercury (Hg)	Radium (Ra)
Arsenic (As)	Copper (Cu)	Molybdenum (Mo)	Strontium (Sr)
Beryllium (Be)	Chromium (Cr)	Nickel (Ni)	Uranium (U)
Cadmium (Cd)	Lead (Pb)	Palladium (Pd)	Zinc (Zn)
Cesium (Cs)	Manganese (Mn)	Platinum (Pt)	

Table 5. Plants Capable of Accumulating Four or More Metals (Degrees of Accumulation: A = Accumulator; H = Hyperaccumulator; P = Precipitator; T = Tolerates)

Scientific name	Common name	Origin and characteristics	Elements	Elements and degree of accumulation
Aquatic/Wetland Plants				
Azolla filiculoides	water-fern	Africa; floating	4	Cu^A; Ni^A; Mn^A; Pb^A
Bacopa monnieri	water hyssop	India; emergent species	5	Cd^H; Cr^H; Cu^H; Hg^A; Pb^A
Eichhornia crassipes	water hyacinth	pantropical/subtropical; troublesome weed	6	Cd^H; Cr^A; Cu^A; Hg^H; Pb^H; Zn^A
Hydrilla verticillata	hydrilla	southern Asia but introduced and spreading as a troublesome weed in the warmer states of the USA	4	Cd^H; Cr^A; Hg^H; Pb^H
Lemna minor	duckweed	native to North America and widespread	4	Cd^H; Cu^H; Pb^H; Zn^A
Pistia stratiotes	water-lettuce	pantropical and native to southern USA but an a aquatic weed	4	Cd^T; Cr^H; CuT; Hg^H
Salvinia molesta	water-fern	India	4	Cr^H; Ni^H; Pb^H; Zn^A
Spirodela polyrrhiza	giant duckweed	native to North America	5	Cd^H; Cr^H; Ni^H; Pb^H; Zn^A
Vallisneria americana	tapegrass	native to Europe and North Africa but widely cultivated in the aquarium trade	4	Cd^H; Cr^A; Cu^H; Pb^H
Crop Plants				
Brassica juncea	Indian mustard		7	Cd^A; Cr^A; Cu^H; Ni^H; Pb^H; Pb^H; U^A; Zn^H
Helianthus annuus	Sunflower		4	Cs^H; Pb^H; Sr^H; U^H; Reference 22 als lists Cd, Cr(VI), Cu, Mn, Ni and Zn absorbed by roots of plants grown hydroponically
Terrestrial Plants				
Agrostis castellana	bentgrass	Portugal	5	Al^A; As^H; Mn^A; Pb^A; Zn^A
Thlaspi caerulescens	alpine pennycress	Europe	7	Cd^H; Co^H; Cr^A; Cu^H; Ni^H; Pb^H; Zn^H
Athyrium yokoscense	fern	Japan	4	Cd^A; Cu^H; Pb^H; Zn^H

For the vascular plants, the database includes both wild and cultivated plant species and varieties. Furthermore, information on biology and ecophysiology, where available, are provided. This includes the origin of the plants, taxonomy, distribution, biology, ecological and health effects, uses of the species, cultivation practices, sources of material for species studies, weedy or other significant relatives, and mode of action in phytoremediation (i.e. tolerance and/or accumulation or hyperaccumulation).

The entries for bacteria, algae, lichens, fungi, and bryophytes with known metal accumulation abilities were included since these organisms act as highly efficient biosorbents that can be used to complement the vascular plants. In particular, sphagnum moss has been documented as an excellent biosorbent [108]. The abilities of mosses and lichens to absorb relatively high concentrations of toxic metals also make them useful as indicators of polluted environments.

In the database, plants were further categorized by their abilities to accumulate, hyperaccumulate, precipitate, or tolerate the specific heavy metals. For the most part, plants were considered accumulators if they contained metal concentrations of 100–200 mg/kg dry weight. Furthermore, species were considered hyperaccumulators if they took up greater than 1000 mg/kg dry weight of most metals, 10000 mg/kg of Mn, or 100 mg/kg of Cd. For the plants, 465 species were found to have some metal remediation capabilites for a single element. Furthermore, 66 species were found with the ability to take up two elements while 25 showed the capacity to accumulate three elements. With the broadest capabilities for heavy metal accumulation, the 15 plant species listed in Table 5 exhibited the ability to accumulate four or more elements. No records were found for beryllium, palladium, platinum, or radium.

Table 6. Number of Records and Plants with the Highest Accumulated Values in PHYTOREM (there are no records for beryllium, platinum, or radium)

Element	Records	Plant with highest recorded concentration	Plant origin	High value (mg/kg dry weight)
Al	25	*Solidago hispida*	Canada	6820
As	4	*Agrostic tenuis (capillaris)*	Cultivated	2000
Cd	37	*Vallisneria spiralis*	India	6242
Co	27	*Haumanistrum robertii*	Africa	10200
Cr	35	*Medicago sativa*	Cultivated	7700
Cs	1	*Helianthus annuus*	Cultivated	High absorbance
Cu	67	*Larrea tridentata*	USA	23700 bioabsorption
Hg	8	*Pistia stratiotes*	Pantropical	1100
Mn	28	*Macdemia neurophylla*	New Caledonia	51800
Mo	1	*Thlaspi caerulescens*	Europe	1500–1800
Ni	372	*Psychotria douarrei*	New Caledonia	4500
		Hybanthus floribundus	Australia	22% total ash
Pb	79	*Brassica juncea*	Cultivated	26200
Sr	1	*Helianthus annuus*	Cultivated	High absorbance
U	3	*Helianthus annuus*	Cultivated	>15000
Zn	48	*Thlaspi caerulescens*	Europe	52000

Furthermore, some of the plants shown in Table 6 were found to be able to take up very high levels of metals. In some cases, the metal concentrations on a dry weight basis were in the relatively high percentage ranges. Some hyperaccumulators such as sunflower and alfalfa represent some important agents with the highest phytoremediation potentials since they are also crop plants with well-established cultivation methods.

5.2 Research Needs and Priorities

For metal-contaminated sites in general, additional research is needed on methods to better identify, quantify, and characterize hazardous substances, the technologies for site remediation, the clean up standards for soil contaminants, the bioavailability of the contaminant, and the significance of health risks posed by impacted sites [105, 106]. Although phytoremediation shows promise as a useful tool for remediation and restoration of sites contaminated by metals, fundamental research and priorities to accelerate the usefulness of these techniques are needed. These include such areas as the basic science behind phytoremediation, the design, implementation, and economics for conducting these techniques, assessments of the ecological and health risks, regulatory acceptance, and biodiversity considerations [3, 4, 20]. Although not mutually exclusive, the specific research needs for each of these areas are listed below.

Basic Science:
- Applicability of phytoremediation to specific contaminants
- Availability, propagation, storage, and protection of plant germplasms
- Effects of soil amendments including synthetic and organic chelators
- Effects of metal speciation and contaminant concentrations
- Improved understanding of forensic botany
- Plant physiology and the basic mechanisms of phytoremediation
- Possibilities and needs for genetic modification of promising plant species
- Rates of uptake and removal
- Selective accumulation and plant screening at impacted sites
- Toxicities of metabolic products
- Understanding of evolutionary engineering and plant adaptation to contaminated sites

Design, Implementation, and Economics:
- Cost data on large-scale projects
- Evaluation of crop rotation for enhancing phytoremediation
- Ideal application protocols for fertilizers, chelating agents, and irrigation
- Improved techniques to predict phytotoxicity
- Individual site testing protocols
- Life-cycle costs
- Screening methods for determining optimal plant species
- Techniques for recovering metals from biomass
- Techniques to minimize site access by herbivores/omnivores

Assessments of Ecological and Health Risks:
- Effects of bioavailability and mobility of the contaminants
- Effects of phytoremediation on the fate and transport of the contaminant
- Fate of chelating agents and chelated metals to partition to the liquid phase
- Fate of volatilized metals
- Impacts to ecological receptors from contaminants taken up into plants
- Risk assessments for comparing different metal remediation technologies
- Toxicity assays for parent and metabolic products

Regulatory Acceptance:
- Models predicting fate and transport effects (both within and outside of plants)
- Multiple site demonstrations under a variety of climatic conditions
- Quality assurance/quality control
- Regulatory oversight for plant disposal alternatives
- Use of bioavailability versus total pollutant loading as regulatory endpoints for contaminated site clean up levels
- Timelines for deployment with respect to regulatory clean up requirements

Biodiversity Considerations:
- Mechanisms to identify and protect promising plant biodiversity at metal contaminated sites
- Mechanisms to allow for equitable distribution of benefits accruing from the use of plant cultivars developed from contaminated sites
- Mechanisms to expand the scope of current biodiversity research to include that found in extreme environments such as contaminated sites
- Protocols for bio-prospecting at contaminated sites
- Mechanisms to evaluate both new plant remediating agents and receiving environments for potential invasiveness

Acknowledgement. This manuscript has not been peer reviewed internally within Environment Canada and as such, reflects only the views of the author. Reference to any particular company or commercial process in this manuscript does not constitute endorsement of said company or product.

5
References

1. Raskin I, Ensley BT (2000) Phytoremediation of Toxic Metals – Using Plants to Clean Up the Environment, Wiley Inter-Science Publication, New York, NY, p ix
2. Vangronsveld J, Cunningham SD (1999) Metal-Contaminated Soils. In Situ Inactivation and Phytorestoration, Springer-Verlag, New York, NY
3. US EPA (2000) Introduction to Phytoremediation. EPA Document # EPA/600/R-99/107, U.S. Environmental Protection Agency Office of Research and Development, Washington, DC
4. US Department of Energy (2000) Proceedings from the Workshop on Phytoremediation of Inorganic Contaminants, Idaho National Engineering and Environmental Laboratory Document #INEEL/EXT-2000–00207, Idaho Falls, ID
5. Fiorenza S, Oubre CL, Ward H (2000) Phytoremediation of Hydrocarbon Contaminated Soil, Lewis Publishers, New York, NY
6. McIntyre TC (1996) A Survey to Determine Public Awareness, Perceptions, and Concerns Over Enhanced Environmental Applications of Biotechnology in Canada. Creative Research Inc., Ottawa, p 36

7. Henry JR (2000) An Overview of the Phytoremediation of Lead and Mercury, United States Environmental Protection Agency Office of Solid Waste and Emergency Response, Washington, DC
8. National Contaminated Sites Working Group (1998) Unpublished Manuscript, Federal Government Working Group on Contaminated Sites, Ottawa, Ontario
9. Chambers C, Winfield M (2000) Mining's Many Faces, Canadian Institute for Environmental Law and Policy, Toronto, Ontario
10. Glass D (2000) The 2000 United States Market for Phytoremediation, D. Glass and Associates, Needham, MA
11. Ernst WH (1998) The Origin and Ecology of Contaminated, Stabilized and Non-Pristine Soils. In: Vangronsveld J, Cunningham SC (eds) Metal Contaminated Soils, In Situ Inactivation and Phytorestoration, Springer, Berlin Heidelberg New York, p 17
12. Seward MR, Richardson, DH (1990) In: Shaw AJ (ed) Heavy Metal Tolerance in Plants:Evolutionary Aspects, CRC Press, Boca Raton, FL, p 75
13. Lasat M (2000) The Use of Plants for the Removal of Toxic Metals from Contaminated Soils, Unpublished Manuscript from the US EPA, American Association for Advancement of Sciences Fellowship Program, p 34
14. Blaylock MJ, Huang JW (2000) Phytoextraction of Metals. In: Raskin I, Ensley B (eds) Phytoremediation of Toxic Metals. Wiley, New York, NY, p 53
15. US EPA (1997) Technology Alternatives for the Remediation of Soils Contaminated with As, Cd, Cr, Hg, and Pb. EPA Document #EPA/540/S-97/500. United States Environmental Protection Agency Office of Emergency and Remedial Response, Washington, DC
16. CCME (1997) Recommended Canadian Soil Quality Guidelines, CCME Document #PN1268. Canadian Council of Ministers of the Environment, Winnipeg, Manitoba
17. Medina VF, McCutcheon SC (1996) Remediation Winter:31
18. Chappell J (1998) Phytoremediation of TCE in Groundwater Using *Populus*, United States Environmental Protection Agency Office of Solid Waste and Emergency Response, Washington, DC
19. National Round Table on Environment and Economy (1998) Brownfields, Ministry of Supply and Services, Ottawa, Ontario
20. McIntyre TC (2001) PHYTOREM – A Global Data Base on Aquatic and Terrestrial Plants Known to Sequester, Accumulate, or Hyperaccumulate Metals in the Environment, Environment Canada, Ottawa, Ontario
21. Glass D (1998) The 1998 United States Market for Phytoremediation, D. Glass and Associates, Inc., Needham, MA
22. Salt D, Blaylock M, Kumar PB, Dushenkov V, Ensley BD, Chet I, Raskin L (1995) Bio/Technology 13:468
23. Saxena P (2001) Unpublished Manuscript on the Establishment of a Network of Centres of Excellence on Phytoremediation in Canada, University of Guelph, Ontario, Canada
24. Chaney RL, Brow SL, Li YM, Angle JS, Homer FA, Green CE (1995) Mining Environ Mgmt 3(3):9
25. Cornish JE, Goldberg WC, Levine RS, Benemann JR (1995) Phytoremediation of Soils Contaminated with Toxic Elements and Radionuclides. In: Hinchee RE, Means JL, Burns DR (eds) Bioremediation of Inorganics, Battelle Press, Columbus, OH, p 55
26. Moffat AS (1995) Science 269:302
27. Brown KS (1995) Bioscience 45:579
28. Kim YH, Yoo YJ, Lee, HY (1995) Biotechnol Letters 17:345
29. Wang TC, Weissman JC, Ramesh G, Varadarajan R, Benemann (1995) Bioremoval of Toxic Elements with Aquatic Plants and Algae. In: Hinchee RE, Anderson DB, Hoeppel RE (eds) Bioremediation of Recalcitrant Organics, Battelle Press, Columbus, OH, p 65
30. Wehrheim B, Weltern M (1994) Appl Microbial Biotechnol 41:725
31. Wilson MW, Edyvean RG (1995) Biosorption for the Removal of Heavy Metals from Industrial Wastewaters. In: Bousher A, Chandra M (eds) Biodeterioration and Biodegrada-

tion 9, Proceedings of the 9th International Biodeterioration and Biodegradation Symposium, Leeds, UK, Sept. 1993, p 47
32. Gardea-Torresdey JL, Bibb J, Tiemann KJ, Gonzalez JH, Arenas, J (1996) Adsorption of Copper Ions from Solution by Heavy Metal Stressed *Larrea tridentata* (Creosote Bush) Biomass, Proceedings of the HSRC/WERC Joint Conference on the Environment, Albuquerque, NM, Great Plains Rocky Mountain Hazardous Substances Research Centre
33. Dushenkov V, Kumar NP, Motto H, and Raskin I (1995) Environ Sci Technol 29:1239
34. Rugh CL, Wilde HD, Stack NM, Thompson DM, Summers AO, Meagher RB (1996) Proc Nat Acad Sci 93:3182
35. Vangronsveld J, Colpaert J, van Tichelen K (1996) Environ Pollut 94:131
36. Schnoor J, Licht LA, McCutcheon SC, Wolfe NL, Carreira LH (1995) Environ Sci Technol 29(7):318A
37. Black H (1995) Innovations 103(12):1106
38. Bosshard PB, Reinhard B, Brandl H (1996) Environ Sci Technol 30:3066
39. Phytotech (1996) The Wall Street Journal, February 29
40. Olson P, Fletcher J (2000) Unpublished Manuscript, Environ Sci Pollution Res
41. Brooks RR (1998) Plants that Hyperaccumulate Heavy Metals – Their Role in Phytoremediation, Microbiology, Archaeology, Mineral Exploration, and Phytomining, CAB International, New York, p 1
42. McCutcheon S (2000) Phytoremediation State of the Science Conference, Boston, MA
43. Gatcliff EG (1994) Remediation 4:343
44. Yaeck J (1997) An Investigation of the Process Required to Determine the Success of Phytoremediation, Unpublished Undergraduate Thesis – Faculty of Science, University of Waterloo, Ontario, Canada
45. Baker AJM, Brooks RR (1989) Biorecovery 1:81
46. Reilley KA, Banks MK, Schwab AP (1996) J Environ Qual 25:212
47. Harrington CF, Roberts DJ, Nickless G (1996) Can J Botany 74:1742
48. Winterhalder K (1996) Environ Rev 4:185
49. Reeves RD, McFarlane RM, Brooks RR (1983) Am J Botany 70:1297
50. Boyd RS, Martens SN (1992) The Raison d'etre for Metal Hyperaccumulation by Plants. In: Baker AJM, Proctor J, Reeves RD (eds) The Ecology of Ultramafic (Serpentine) Soils, Intercept, Andover, p 279
51. Pollard JA, Baker AJM (1997) New Phytol 135:655
52. Borovik AS (1990) Characterization of Metal Ions in Biological Systems. In: Shaw AJ (ed) Heavy Metal Tolerance in Plants: Evolutionary Aspects, CRC Press, Boca Raton, FL, p 3
53. US Department of Energy (1994) Summary Report of a Workshop on Phytoremediation Research Needs, Office of Technical Information, Springfield, VA
54. Sorenson DL, Sims RC, Qiu X (1994) Field Scale Evaluation of Grass Enhanced Bioremediation of PAH Contaminated Soils. 20th Annual RREL Research Symposium Abstract Proceedings, EPA Document #EPA/600/R-94/011, United States Environmental Protection Agency, Washington, DC, p 92
55. Schwab AP, Banks MK, Arunachalam M (1995) Biodegradation of Polycyclic Aromatic Hydrocarbons in Rhizosphere Soil. In: Hinchee RE, Anderson DB, Hoeppel RE (eds) Bioremediation of Recalcitrant Organics, Battelle Press, Columbus, OH
56. Cunningham SD, Berti W, Huang JW (1995) Remediation of Contaminated Soils and Sludges by Green Plants. In: Leeson A, Alleman BC (eds) Phytoremediation Proceedings of the 3rd International Conference on In Situ and On-Site Bioremediation, San Diego, CA, p 33
57. Anderson TA, Guthrie EA, Wilson BT (1993) Environ Sci Technol 27:2630
58. Anderson TA, Kruger EL, Coats JR (1994) Chemosphere 28:1551
59. Berti WR, Cunningham SD (1997) Environ Sci Technol 31:1359
60. Kramer U, Cotter-Howells JD, Charnock JM, Baker AJM, Smith JAC (1996) Nature 379:635

61. Burken JG, Schnoor JL (1996) J Environ Engrg 122:958
62. Brooks RR, Morrison RR, Reeves RS, Malaisse F (1978) Plant and Soil 50:503
63. Brooks RR, Shaw S, Asensi MA (1981) Physiol Plant 51:167
64. Van Der Lelie D, Schwitzguebel JP, Glass D, Vangrosveld J, Baker A (2001) Environ Sci Technol, November 1, p 447A
65. Coghlan A (1996) New Scientist 17:17
66. Ruesink JL, Parker IM, Groom MJ, Kareiva PM (1996) BioScience 45(7):465
67. Office of Technology Assessment (1993) Harmful Non-Indigenous Species in the United States, US Government Printing Office Document #OTA-F-565, Washington, DC, p 198
68. Keddy C (2000) The Role of Importation Control in Protecting Native Canadian Biodiversity – A Discussion Paper, Canadian Wildlife Service, Environment Canada, Ottawa, Ontario, p 9
69. Small E (1993) Can J Botany 71:1537
70. Keeler K, Turner C (1991) Management of Transgenic Plants in the Environment. In: Levin M, Strauss H (eds) Risk Assessment in Genetic Engineering:Environmental Release of Organisms, McGraw-Hill Inc, New York, NY
71. Lancaster J (1997) Alberta Native Plant Council Guidelines for Rare Plant Surveys, Alberta Native Plant Council, Alberta, Canada
72. Mack RN (1996) Biol Conservation 78:107
73. Tilman D (1997) Ecology 78(1):81
74. Fowler BA (1992) Environ Health Perspect 100:57
75. Leffler PE, Nyholm NE (1996) Ambio 25:417
76. Leggett RW (1989) Health Phys 57:365
77. Stein JH, Gottschall J, Osgood RW (1975) Kidney Int 8:27
78. Lock EA (1993) Responses of the Kidney to Toxic Compounds. In: Ballantyne B, Marrs T Turner P (eds), General and Applied Toxicology, vol 1. MacMillan, New York NY
79. Luckey TD, Venugopal B (1977) Metal Toxicity in Mammals. Plenum Press, New York, NY
80. Seiler HG, Sigel H, (1988) Handbook on Toxicity of Inorganic Compounds. Marcel Dekker, New York, NY
81. Babich H, Devanas MA, Stotkzy G (1985) Environ Res 37:253
82. O'Brien P, Kortenkamp A (1994) Environ Health Perspect 102(3):3
83. Kawanishi S, Inoue S, Yamamoto K (1994) Environ. Health Perspect 102:17–20
84. Stohs SJ, Bagchi D (1995) Free Rad Biol Medic 18:321–336
85. Arhar M, Abdulla M, Sultana S, Favier A (1993) J. Trace Elem Exper Med 6:65–73
86. Domingo, JL (1994) J Toxicol Environ Health 42:123–141
87. Bauchinger M, Schmid E, Einbrodt, J (1976) Mutat Res. 40:57–62
88. Anderson SL, Wild GC (1994) EnvironHealth Perspect 102:9–12
89. Hebert PD, Luiker MM (1996) Sci Total Environ 191:23–58
90. McBee K, Bickham JW (1998) Bull. Environ Contam Toxicol 40:343–349
91. Sileo L, Beyer WN (1985) J Wild Dis 21:289
92. Wren CD (1986) Environ Res 40:210
93. Sivertsen T, Daae HL, Godal A, Sand G (1995) Environ Pollut 90:75
94. Cloutier NR, Clulow TP, Lim TP, Dave NK (1985) Environ Pollut (Series B) 10:19
95. Hunter BA, Johnson MS (1982) Oikos 38:108
96. Sample BE, Opresko DM, Suter II GW (1996) Toxicological Benchmarks for Wildlife: 1996 Revision, US Department of Energy Document #ES/ER/TM-86/R-3, Oak Ridge National Laboratory, Oak Ridge, TN
97. Sample BE, Aplin MS, Efroymson RA, Suter II GW (1997) Methods and Tools for Estimation of the Exposure of Terrestrial Wildlife to Contaminants, US Department of Energy Document #ORNL/TM-13391, Oak Ridge National Laboratory, Oak Ridge, TN
98. Glooschenko V, Downes C, Frank R (1988) Sci Total Environ 71:173
99. Kronberg B, Glooschenko V (1994) Alces 30:71
100. Charters DW, Christian JJ (1987) Indigenous Mammals as Indicators of Environmental Pollution and Potential Health Risk at Chemical Dumps. In: Saxena J (ed)

Hazard Assessment of Chemicals, Vol. 5, Hemisphere Publishing Corporation, Washington, DC
101. McBee K, Lochmiller RL (1996) Ecotoxicity and Human Health, CRC Press, Boca Raton, FL
102. O'Brien DJ, Kaneene JB, Poppenga RH (1993) Environ Health Perspect 99:351
103. Carson B, Ellis HV, McCann JL (1986) Toxicology and Biological Monitoring of Metals in Humans, Lewis Publishers, Inc., Chelsea, MI
104. Gardea-Torresdey JL, Tang L, Salvador JM (1996) J Hazard Mat 48:191
105. Hrudy S, Pollard SJ (1993) Environ Rev 1:55
106. Alexander M (1999) Biodegradation and Bioremediation, 2nd edn. Academic Press, New York, NY

Received: June 2002

Maintaining Hydraulic Control Using Deep Rooted Tree Systems

A. Ferro[1] · M. Gefell[2] · R. Kjelgren[3] · D. S. Lipson[2] · N. Zollinger[1] · S. Jackson[1]

[1] Phytokinetics, Inc., 1770 N. Research Park Way, Suite 110, North Logan, UT 84341, USA. *E-mail: ariferro@phytokinetics.com*

[2] Blasland, Bouck & Lee, Inc., 6723 Towpath Road, P.O. Box 66, Syracuse, NY 13214-0066, USA. *E-mail: MJG@BBL-Inc.com*

[3] Department of Plants, Soils, and Biomet., 4820 Old Main Hill, Utah State University, Logan, UT 84322-4820, USA. *E-mail: Rkjel@mendel.usu.edu*

Groundwater phytotechnology systems are currently being used to control the migration of contaminant plumes and remove groundwater that flows into containment areas. Trees in such systems function by reducing recharge and using groundwater via transpiration. A stand of deep-rooted trees can act as a biological "pump", removing a substantial amount of water from the saturated zone. Depth to groundwater compared to rooting depth is an important consideration. Many useful tree species are not naturally deep-rooted, but various cultural practices can produce trees with roots that are deep enough to tap into the water table. Hydraulic control can be calculated or modeled using equations that compare transpired water to groundwater flow and take into account site-specific geologic and climatic variables. However, field measurements should be carried out whenever possible to obtain more precise estimates of groundwater capture and control.

Keywords. Groundwater, Hydraulic Control, Plume Control, Phytotechnology, Phytoremediation, Capture Zone

Advances in Biochemical Engineering/
Biotechnology, Vol. 78
Managing Editor: T. Scheper

Abbreviations and Symbols

A	Area
a	$(K_r(K_z)^{-1})^{0.5}$
AEM	Analytic element method
B	Aquifer thickness
bgs	Below ground surface
ET	Evapotranspiration
ET_o	Reference evapotranspiration
Fs	Sap flow
I	Hydraulic gradient

K	Hydraulic conductivity
K_c	Water loss coefficient
K_r	Radial, horizontal hydraulic conductivity
K_z	Vertical hydraulic conductivity
LA	Leaf area
LAI	Leaf area index
MTBE	Methyl-*tert*-butyl-ether
PM	Penman-Monteith
Q	Groundwater flow
RH	Relative humidity
R_n	Solar radiation
T_a	Ambient air temperature
TBA	*Tert*-butyl alcohol
TDP	Thermal dissipation probe
U	Wind Speed
USGS	United States Geological Survey
V_T	Volumetric water use
W_o	Width of capture zone
W_{max}	Maximum width of capture zone
X_{stag}	Stagnation point
Z_{max}	Maximum depth of capture zone
Z_o	Capture zone depth
δD	Molar ratio of deuterium to hydrogen
θ	Water use multiplier
Ψ	Water potential

1 Introduction

Groundwater phytotechnology systems are currently being evaluated at many sites, primarily in the United States. These systems are frequently dense stands of water-loving trees that are grown using cultural practices that encourage the development of deep roots and use groundwater as a primary source of moisture. Groundwater phytotechnologies are applicable for sites with dissolved-phase organic chemical contaminants present at a wide range of concentrations.

1.1 Environmental Protection Objectives

In general, the environmental protection objectives for contaminated site are containment, remediation, or both. When used strictly for containment, the use of trees for hydraulic control is included in the broader area of phytotechnologies. On the other hand, when used to promote the degradation or reduction of contaminant concentrations, the use of trees can be referred to as phytoremediation, a specific area within phytotechnologies.

In the context of phytotechnologies, hydraulic control is defined as the use of plants to remove groundwater in order to contain or control contaminant migration in the subsurface. Zones of contaminated groundwater, or "plumes," migrate downgradient through a geologic medium, favoring the most permeable pathways. The goal of a groundwater phytotechnology system is to remove as much water as possible from the plume with a dense stand of deep-rooted phreatophytic trees (trees that use water directly from the saturated zone) and ultimately prevent the plume from migrating to downgradient receptors. Transpirational water use by trees has two important effects on the local water budget: groundwater removal and recharge reduction. Both contribute (directly or indirectly) to groundwater containment and capture by trees.

1.1.1 *Remediating Contaminant Plumes*

Deep-rooted trees may be installed in dense multiple rows at the leading edge of the plume with the rows set perpendicular to the direction of groundwater flow. The trees will transpire groundwater at a substantial rate, depending on climatic factors and the age of the stand. This natural water removal will create a zone of capture where all of the groundwater within a certain thickness of the saturated zone is used by the trees. Contaminants dissolved in the groundwater that enter the capture zone are drawn into the rhizosphere of the trees. The fate of the contaminants is highly compound specific, but may include biodegradation in the rhizosphere, plant uptake followed by metabolic transformation or volatilization, or immobilization in the root zone. The extent to which a phytoremediation system decreases dissolved groundwater contaminants can be assessed by analyzing water samples from monitoring wells.

1.1.2 *Groundwater Removal*

Another common remedial objective is to hydraulically control groundwater that flows into a containment area such as into an area behind a barrier wall. Because a dense planting of deep-rooted trees can remove groundwater at a certain rate, plant transpiration can be used in lieu of mechanical groundwater pumping, at least on a seasonal basis. At many sites, contaminated water (groundwater or leachate) appears on the surface as seeps. Dense multiple rows of deep-rooted trees can be installed upgradient of the seeps with the objective of partially de-watering the saturated zone. Depending on the hydrogeology of the site, planting density, and climate, a stand of trees may be able to dry up the seeps.

1.2 Use of Trees for Environmental Protection

Rapidly growing phreatophytic trees with large crowns and high transpiration rates are well suited as biological pumps. These include riparian species such as poplars (*Populus*) and willows (*Salix*) in cool temperate regions and eucalypts (*Eucalyptus*) in warm temperate regions. These species tend to have less stomatal control under conditions of high evapotranspiration (ET) demands (see Chapter 2). A significant portion of the water used by phreatophytes is presumably derived from the zone immediately above the water table known as the capillary fringe. This zone is the interface between the saturated and unsaturated soils. Transpirational removal of water from the capillary fringe causes an upward wicking of groundwater from the saturated zone. Deep-rooted phreatophytic trees can use groundwater at a rate sufficient to influence flow patterns and hydraulically control or capture contaminant plumes.

1.2.1
Root Development

A vigorous root system is important for tree growth because the major portion of water and mineral nutrients absorbed enters through the roots. An extensive root system enables trees to adapt to a wide range of soil conditions and to compete with other plants. Soil water potential is one of the main factors affecting root growth. At the extreme, root growth becomes severely inhibited at values below about -0.7 MPa [1]. The capillary movement of water from wet soil to dry soil is extremely slow when soil moisture drops below field capacity, so continuous root extension is necessary for water absorption. Other environmental factors that affect root development include soil temperature, subsurface oxygen concentration, soil bulk density, and water table depth.

Roots respire and therefore must have a constant supply of oxygen to remain healthy. Under conditions where O_2 is limited or becomes depleted, root growth is reduced. An ideal rooting medium is more than 18% O_2 and less than 2% CO_2 [2]. Flooding decreases soil aeration because water displaces air from the soil pores. For most species, flooding causes root growth to stop and the existing root system to die and decay [3, 4]. Roots are more sensitive to flooding during the growing season than when trees are dormant. This is due to the high metabolic demand placed on the roots during the summer months when transpirational water use is high [5].

Some species can adapt morphologically and physiologically to flooded soils [6]. Morphological adaptations in flood tolerant trees allow the transfer of oxygen from aerial tissues, through the trunk, to the roots, and out into the rhizosphere [7, 8]. Uptake of oxygen by aerial tissues is facilitated by lenticels, spongy areas on the stem that serve as pores for gas exchange. Oxygen uptake from aerial tissues and internal transport has been demonstrated for *Populus* and *Salix* species, common phreatophytes [9–11].

Soil compaction also decreases soil aeration and adversely affects the environment of the root zone. Soil macropores greater than 50 µm, which normally function as air and water spaces, are lost in compacted soil. Compaction decreases the amount of soil water that is available for plant uptake and decreases the infiltration of rainwater [12]. Compacted soil also reduces root penetration. For example, root penetration by Austrian pine seedlings is restricted at a bulk density of 1.4 to 1.6 g cm^{-3} [13].

1.2.2
Rooting Depth

Rooting depth is an important consideration for groundwater phytotechnologies. Many useful tree species are not naturally deep-rooted, but special cultural practices can promote unusually deep root development. Roots proliferate in the heterogeneous soil volume that is most favorable in terms of air, water, and nutrient supply [14–16]. The upper portion of the soil profile supports an intensive growth of tree roots because of adequate rates of gas diffusion for root respiration, higher nutrient availability, and lower soil bulk density. Consequently, most tree roots grow horizontally rather than vertically [17]. Thus, many root distribution patterns are due to these heterogeneous environmental conditions rather than the genetic tendencies of the species [18]. For example, trees in the tropical forests of Brazil have a mean maximum rooting depth of 4 to 9 m, but where soil conditions permit, roots 18 m deep have been observed [19]. Similarly, fir trees (*Abies*) can produce dominant taproots that are 4 m or longer [20]. However, where soil conditions were not conducive to deep roots, fir trees developed only a very shallow taproot (<0.5 m) [21]. This type of phenotypic plasticity is a particularly important trait when attempting to "engineer" deep-rooted trees that naturally tend to be shallow-rooted.

Although fewer roots are present in the deeper portions of the soil profile, these roots may serve a disproportionately important function. Deep roots can provide the plant with water when moisture levels are low in the upper portions of the profile coupled with high transpirational demands. A relatively low density of roots is required to withdraw water from the water table (or capillary fringe) compared to the intensive concentration of roots involved in nutrient and water absorption near the surface. Groundwater has relatively high water potentials. Consequently, it is extracted more efficiently than water from the vadose zone. For example, 20% of the water transpired by winter wheat during dry periods can be taken up by deep roots that amount to only 3% of the total root weight [22].

Some plant species have roots that can naturally reach impressive depths. Phreatophytic plants can have extremely deep roots because they depend on groundwater, which is sometimes very deep in the soil profile. *Boscia albitrunca*, a Kalahari Desert plant, has been shown to have root depths of up to 68 m [23]. Similarly, *Juniperus* roots have been seen in mine shafts at depths greater than 61 m [20]. Figure 1 shows the average maximum as well as the maximum rooting depths for a variety of tree genera.

For many species, rooting depth can be correlated with the depth to the saturated zone. Furthermore, many species with shallow, spreading root systems grow in areas where the groundwater levels are close to the surface [25]. However, roots can be lured to greater depths using special planting methods. On dry, upland sites, where the saturated zone is less than 2.4 m below ground surface (bgs), boreholes can be drilled and long hardwood cuttings (2.7 m long "poles") can be planted directly into the moist sediments. Each borehole can then be backfilled with a porous rooting medium, leaving about 0.3 m of each pole extending above the ground surface. Roots develop along the length of the pole that is exposed to

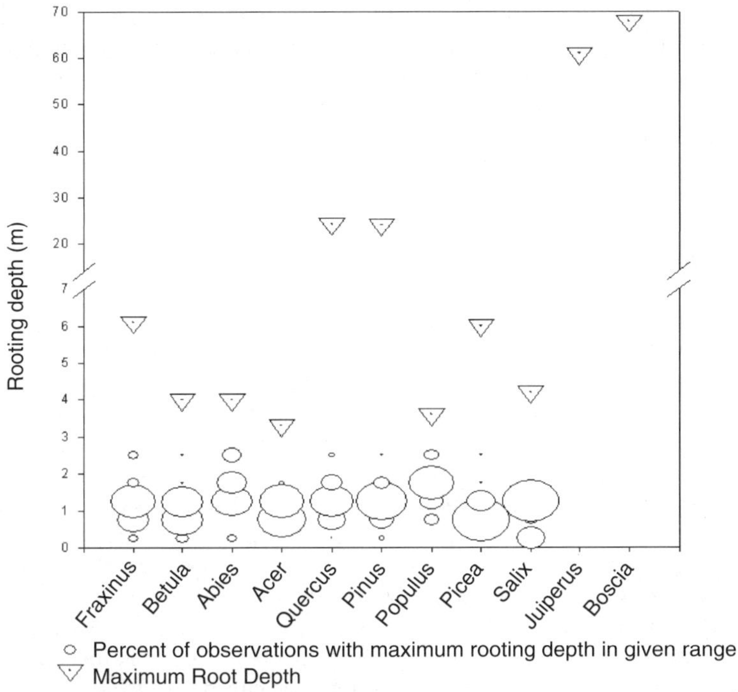

Fig. 1. Average Maximum and Maximum Rooting Depths of Several Tree Genera. The data shown as circles are maximum rooting depths found in one study [24] and represent the percent of observations with maximum rooting depths within a given range. These data were collected by examining root plates of trees in England blown over during a storm in 1987. The total area of the circles for a given genus equals one-hundred percent of the observations. The maximum rooting depth data, shown as triangles, have been selected from literature reviews of maximum rooting depths [20, 22]

moist sediments. No surface irrigation is necessary even at extremely hot, arid sites because the trees are immediately dependent upon groundwater for moisture. The "deep-rooted" trees can grow rapidly, up to 3 m per year. Recently, the root system of a 3-year-old poplar tree planted using these deep rooting techniques was investigated [26]. An analysis indicated that roots had grown out of the borehole and proliferated throughout the vadose zone. However, a few feet above the saturated zone, the roots were heavily concentrated in the borehole and extended down into the water table.

1.2.3
Suitable Site Characteristics

Hydraulic control of groundwater is generally feasible at sites where the water table is within 3 to 4 m of the ground surface and contains a relatively low volumetric flow rate through a given cross section of the plume. These conditions exist where the plume is vertically thin, the hydraulic gradient is relatively low, and/or the geologic medium has low to moderate permeability. Plumes in the upper portion of the saturated zone are more amenable to phytotechnologies than deeper plumes. Generally, strategies for hydraulic control using trees are more feasible in arid regions where the recharge rate is limited and abundant sunlight and high temperatures enhance the rate of transpiration.

1.2.4
Phytotoxicity

As with other phytotechnology applications, phytotoxicity is a limiting factor. Certain concentrations of dissolved organic chemicals in the groundwater are toxic to trees. For example, a zone of light non-aqueous phase liquid (LNAPL) on the upper surface of a water table can be toxic to trees and can prevent the roots from tapping into the saturated zone. Certain tree species can exhibit a higher tolerance to organic contaminants than other species. At a given concentration, certain organic compounds are also more toxic than others. Therefore, greenhouse phytotoxicity studies are an important first step in designing groundwater phytotechnology systems.

In addition to the potential phytotoxicities of the contaminants, other soil properties can limit or inhibit the growth of the vegetation. These can include the groundwater salinity, pH, and nutrient content. Furthermore, the presence of soil pathogens and other pests and diseases can be limiting factors at some sites. In these cases, tree species with specific or a variety of tolerances must be used.

2
Transpirational Water Use

Tree roots take up water from the soil. En route to the leaf through the xylem vessels, a small amount of this water is used for growth. However, the bulk of the water is lost through transpiration, which is the evaporation of water from inside a plant leaf with the subsequent movement of the water vapor into the ambient air. The transpirational stream of water from plant roots to the leaf is under tension (expressed as water potential, Ψ) where the higher the transpiration rate, the greater the tension. Air temperature (T_a), relative humidity (RH), solar radiation (R_n) absorbed by the leaf, and wind (U) are atmospheric factors that drive transpiration. In moist soil, the rate of transpiration is largely a function of the difference in vapor pressure between the mesophyll of the leaf and the atmosphere.

2.1
General Background

All types of vegetation transpire water obtained from the soil either from infiltration of surface water or groundwater. The ability of most temperate plants to regulate this process occurs primarily through pores known as stomates. However, different types of vegetation have different degrees of control of this process.

2.1.1
Stomatal Regulation

Most vegetation lose water vapor primarily through the stomata, the tiny controllable pores generally on the underside of the leaf. These pores regulate transpiration and allow gas exchange during photosynthesis. The stomata occupy about 1% of the epidermal surface. The remainder of the leaf surface is more resistant to water loss, and in many species, is covered with a waxy cuticle. Diffusion of water vapor through the stomata of a leaf can exceed 50% of the rate of evaporation from a free water surface with the same area as the leaf. Stomatal pores are very small (10 to 20 μm in length and 5 to 10 μm wide when open) and numerous (13000 to 100000 per cm^2). The size of the stomatal opening is controlled by two guard cells. The pores increase in size as the turgor of the guard cells increase and vice versa. The stomata open in response to both the presence of light and low concentrations of CO_2, but close in darkness and when water stress causes a loss of turgor. For many plants, stomata also close in response to dry air and the shaking caused by wind.

2.1.2
Herbaceous Species

Species vary in how they regulate transpirational water loss. Low-growing, densely spaced herbaceous plants, such as typical crop species, generally have high transpiration rates. The distance water travels from roots to leaves is short enough that internal plant water does not come under substantial tension. The dense spacing of the plants creates a relatively thick layer of still air (the boundary layer) saturated with water vapor that buffers the stomata from the humidity of the ambient air. Consequently, the stomata of such herbaceous plants exercise only partial control over transpiration with wind speed and radiation largely determining the transpiration rate instead.

2.1.3
Woody Plants

Woody plants regulate transpiration differently than herbaceous species because of their greater height and ventilation. Stomatal sensitivity moderates tree transpiration [27], particularly when trees are well ventilated. Specifically, wind moves ambient air closer to tree leaves than a dense stand of herbaceous plants. Therefore, the boundary layer around tree leaves is small compared to a low, dense stand of plants. Furthermore, greater height results in greater ventilation, especially for trees that are essentially isolated or are spaced far enough apart to preclude a closed forest canopy. The resulting minimal boundary layer means that stomates are not buffered from wind and thus exert maximum control over transpiration. Isolated trees usually have higher transpiration rates compared to a single tree in a closed canopy because an individual tree crown has a higher percentage of its crown illuminated and ventilated. In addition, the distance water travels from the roots to the leaves is generally much greater in trees than herbaceous plants. Therefore, the water column is under greater tension. These two factors result in the stomata of trees partially closing at midday in order to moderate transpiration and internal water tension.

Tree transpiration rates can be very high in the morning hours, especially in the summer, when evaporative demand is low due to high humidity and low air temperature. Water transpired during the course of a day is lost initially from internal tree storage, resulting in increased water tension or internal water deficit. The decrease in Ψ is greatest at the sites of transpiration in the leaf, causing water to flow from the xylem into the leaf along a Ψ gradient. In turn, the ascent of the xylem sap leads to a reduction in the water potential in the roots, so that water ultimately flows from the soil into the roots later in the day [1]. Under high evaporative demand, midday tree Ψ can fall low enough that water supply does not keep up, even with adequate soil moisture. This results in partial closure of the stomata to moderate demand and alleviate internal water deficit, but the stomatal regulation can also reduce CO_2 uptake and photosynthesis. In moist soil, the water deficit is usually made up by late afternoon or at night. In drier soil, the internal water deficit increases more rapidly during the day, so that leaf turgor falls, and stomata close earlier in the day or fail to open. Water stress can occur as the soil water content decreases during a drought and water movement to the roots becomes too slow to replace the water lost during the previous day. Roots are the last tissues to undergo water stress and are also the first to recover because they are closest to the water source.

Within a genus, species from similar habitats have similar transpiration rates [28]. This suggests similar stomatal behavior. However, substantial variation in stomatal behavior exists between tree types. Woody species from arid habitats may have sunken stomates, leaf hairs, and other morphological characteristics that limit overall stomatal conductance. In many tree species, stomates are sensitive to low humidity, particularly in conifers and trees native to arid regions [29].

2.2 Groundwater Use by Deep-Rooted Trees

Although trees have roots that extend into the saturated zone, the trees may or may not use groundwater as their primary source of water. However, trees that are tapping into the groundwater have been shown to have profound effects on the water table elevations beneath the stands of trees.

[1] Water is used first from the leaves and sapwood because resistance to water removal from adjacent tissues is lower than the resistance to uptake of water from the soil. Water stored in the xylem, therefore, provides the plant with a certain capacitance, and midday shrinkage of plant tissue can occur.

2.2.1
Water Utilization

An assessment of water balance is clearly of central importance in the design of groundwater phytotechnology systems using deep-rooted trees. The groundwater removal rate cannot be greater than the rate of transpiration and is probably less because some portion of the transpired water is derived from precipitation. Making the assumption that the trees are opportunistic and use water derived from precipitation whenever it is available, then the rate of groundwater use during a given time period should be approximately equal to the total rate of water use minus the vadose zone moisture that is available for plant uptake.

Studies indicate that some trees have the ability to switch rapidly between different water sources as they become available for plant uptake. For instance, as soils dry, riparian trees of the Sierra Nevada shift from using soil moisture to using groundwater [30]. In eastern forests, *Pinus strobes* shifted from using water in deeper soil layers to surface soils following precipitation [31]. Similarly, *Tamarix* was also observed to have the ability to switch from the use of groundwater to surface water after a precipitation event [32].

Some studies show exclusive groundwater use by certain trees, regardless of the availability of surface water. Junipers growing on sites with regular summer precipitation utilize water from the unsaturated zone immediately following rain. However, those growing on sites with lower summer precipitation rates do not respond to the rain showers [33]. Riparian *Quercus* and *Acer* in the Intermountain West were observed to shift from surface to deeper water sources as the trees matured. Furthermore, they used little or no water from the adjacent perennial stream at maturity [34].

A variety of methods have been devised to minimize the percentage of transpirational water derived from precipitation. Planting grass or diverting water with soil mounds can reduce infiltration at the base of the tree. One specialized approach to maximize water uptake from the saturated zone involves planting trees in long casings or tubes, which prevent the roots from freely exploring the vadose zone [35].

2.2.2
Effects on Water Table Elevations

Transpirational water use by deep-rooted trees can lead to a decrease in water table elevation, which is a result of both groundwater removal and reduced recharge. Specifically, several researchers have observed diurnal water table fluctuations under stands of trees. Diurnal fluctuations were noted under a stand of mesquite trees (*Prospopis juliflora*) in southeastern Arizona [36]. The 0.23 ha stand caused water table fluctuations that frequently exceeded 30 mm per day. Water table elevations declined as transpiration increased at mid-day and recovered somewhat when transpiration rates decreased at night. Similarly, daily water table fluctuations were reported under a dense stand of willow trees (*Salix*) in Escalante Valley, Utah [37]. The average daily fluctuation of 9.5 cm was attributed to transpirational water use from the saturated zone, which was 1.5 to 1.8 m bgs.

The effects of transpirational water use on water table elevations are also evidenced by the rise in the water levels after cutting. In 1949, a loblolly pine-hardwood forest in North Carolina was thinned to a density of 182 trees per ha [38]. Water table elevations were compared in the thinned area with an adjacent uncut area of forest with a density of 209 trees per ha. The groundwater beneath the thinned section reportedly rose 2.25 m higher than groundwater in the uncut area after approximately 2 years. In July 1952, the uncut area was cut clear in strips 60 m wide, extending the length of the study area. In the month following the cutting, the water table rose 2.68 m in the cut cleared areas but only 0.12 m in the thinned area. By the end of the year, the water table in the cut cleared areas was 2.13 m higher than the water table in the area that was thinned. It was concluded that the water table rise was largely a result of the loss of transpirational water use in the cut areas [38].

Conversely, reforestation has been shown to cause a decline in the water table elevation [39]. Groundwater levels beneath newly planted pine (75 to 225 trees per ha) and eucalyptus (150 to 625 trees per ha) stands in Western Australia were compared to groundwater levels beneath an adjacent pasture [40]. In the first 10 years, groundwater declined 1.0 m under the pine stand relative to the pasture. The Eucalyptus system lowered the water table 2.0 m compared to the pasture over 7 years. It was further noted that additional monitoring must be done to determine what factors contributed the most to the decline.

2.3 Measuring and Estimating Water Use by Trees

Quantifying water use is critical in determining the effectiveness of trees in controlling groundwater. While methods exist to directly measure tree water use, these empirical methods are generally costly and labor intensive, making them impractical on large-scale and day-to-day levels. A practical alternative is to estimate water use from easily measurable climatic and physiological factors. In order to determine whether the primary source of water for a tree originates from groundwater or vadose zone water derived from a recent precipitation, xylem water can be analyzed and compared to the two potential sources.

2.3.1 *Reference Evapotranspiration*

Agricultural scientists have developed methods for measuring water loss (evaporation-transpiration, or ET) under standardized conditions. Water loss measured under these standardized conditions at one time and place can therefore be compared to water loss measured under the same conditions (but different climatic parameters) at other times and places. Historically, reference evaporation was measured from a pan of water (evaporation pan), but difficulties in maintaining uniform measurements and conditions combined with the advent of electronic weather stations, have reduced the use of evaporation pans.

Several empirical equations correlating air temperature to evapotranspiration have been used extensively [41], but these equations were not particularly robust in a wide variety of climates. The most accurate method for estimating plant water loss is based on the energy balance of a vegetated surface as described in the Penman-Monteith (PM) equation [42]. The major parameters used in the equation are the climatic factors T_a, RH, R_n, and U. The PM equation, as it is currently constituted, represents reference evapotranspiration (ET_0) from a hypothetical 12 cm high cool-season grass stand where plant density and stomatal behavior are held constant.

2.3.2
Water Loss Coefficients

Plant factors that affect transpiration rates vary from species to species, and while water loss of a particular species is proportional to a reference ET, it is rarely the same as ET. Plant control over water loss varies among species with the interrelationship of stomatal behavior, canopy density, leaf arrangement, and plant height. Densely spaced, low growing crops essentially present a "big leaf" to the atmosphere in which the stomata are buffered from ambient climatic conditions by the boundary layer above the canopy. Dense plant stands with high leaf area indices (LAI, the leaf area per unit area of ground surface) have high transpiration rates in the sunlit top layer [43]. Transpiration rates decrease rapidly in the shaded, lower layers of leaves where the stomata are more closed and there is less energy for evaporation. Leaf orientation can also affect transpiration. Vertically oriented leaves, such as a dense grass canopy, allow for greater light penetration and illumination of a greater percentage of the leaf area, which, in turn, results in a potentially higher rate of transpiration.

A species-specific water loss coefficient, K_c, was developed to address variations in plant water loss. K_c equals the empirical plant water loss divided by the reference water loss and is functionally related to the reference ET such that for a given species the plant factors that regulate transpiration (height, density, stomatal behavior) are assumed to be constant over an expected range of reference ET values in different climates and locations. The type of reference ET used when developing a K_c will affect how that factor can be used. A K_c value developed using pan evaporation as a reference, will underestimate actual plant water loss if used in conjunction with ET_0 because pan evaporation is about 1.2-fold higher than ET_0. Therefore, it is important that the current PM equation be as widely adopted as possible for calculations of ET so that plant water loss estimates are comparable.

Using K_c values to estimate plant transpirational water use for woody plants is more difficult than for low growing crop plants. Although accurate crop coefficients have been developed to estimate fruit tree water use in uniform orchards [44], estimating tree water use in other settings, such as urban landscapes or phytotechnology sites, is complicated by species diversity and tree spacing. Another complicating factor in using water loss coefficients for woody plants is the stomatal sensitivity to humidity. This variation in stomatal behavior violates the basic assumption implicit in estimating K_c that plant factors remain constant over a range of expected ET values. Variation in stomatal sensitivity to humidity may be small for some tree types such as temperate broadleaf deciduous species. For these species, an approximate K_c factor can be developed. However, some woody species such as plants native to dry habitats may close their stomates during periods of high ET [45], greatly reducing the accuracy of K_c. Another problem with applying a K_c to woody plants is that water loss is three-dimensional and occurs with both vertical and lateral wind movement. Values for K_c developed for low growing crop plants imply only vertical water loss. While transpiration for an isolated tree can be averaged over the crown to arrive at a two-dimensional water loss value, this calculation is imperfect because the rate of water loss from the isolated tree can change with age and crown density. Finally, the development of a K_c that represents transpiration of an isolated tree will represent only the tree crown and will not take into account surface evaporation. These limitations do not preclude developing a relationship between K_c and ET, but users must recognize that such a relationship would create a coefficient that is only loosely proportional and not functionally related to ET. For most woody species, the water use multiplier (θ, see Section 2.3.3) is less accurate than K_c and only represents tree water loss (i.e. it does not include surface evaporation). Therefore, θ should be used carefully depending on the conditions surrounding the tree to which it is being applied.

2.3.3
Water Use Models

Transpirational water use for an isolated tree during a given time period can be estimated using Eq. (1).

$$V_T = ET_o \times \theta_f \times LA \tag{1}$$

In Eq. (1), V_T is the total water use in units of volume per tree, ET_o is the reference evapotranspiration during the time period, θ_f is the water use multiplier for an isolated (freestanding) tree, which is equivalent to the rate of water use per leaf as a percentage of ET_o, and LA is the total leaf area of the tree.

Values for V_T for a stand of trees can be estimated using Eq. (2).

$$V_T = ET_o \times \theta_s \times LAI \times A \tag{2}$$

In Eq. (2), θ_s is the water use multiplier for a tree within a stand, LAI is the leaf area index, and A is the area of the stand of trees.

For a stand of trees, values for θ_s and LAI are species-specific and depend on planting density and stand maturity. For example, canopy closure for a dense stand of poplar trees generally occurs during the fifth growing season, at which time LAI attains a maximal steady state value. Before canopy closure, the trees can be considered isolated from one another. Values for the water use multiplier (θ) are higher for these freestanding trees (θ_f) than for trees within a stand (θ_s). At the time of canopy closure, θ values decrease primarily because of self-shading.

Equations (1) and (2) can give reasonably accurate estimates of pumping capacity. However, another key factor in estimating V_T for groundwater phytotechnology systems is potential plant stresses. The equations presented above assume that the trees are freely transpiring. Any kind of stress – water, nutritional, toxic substances, etc. – can reduce photosynthesis and stomatal conductance. If stomates close, the transpiration rate is directly reduced, and a water use multiplier cannot be applied to the stand. Another factor that must be considered is stand density. Water use by a stand of trees that has not reached canopy closure is calculated as the sum of all the freestanding trees in the stand using Eq. (1). At canopy closure the stand essentially becomes a "big leaf," similar to a low-growing crop canopy. In this situation, pumping capacity can be estimated using Eq. (2). Leaf illumination and ventilation is higher at the beginning of crown closure because more of the crown area of an individual tree is exposed to sunlight. Under these conditions, there are more open stomates, more energy for evaporation, and more air movement than for a closed canopy. On the other hand, in a closed canopy, many of the lower leaves senesce and drop due to insufficient light. The combination of high individual tree water loss for closely packed but not overlapping stand of trees theoretically results in a higher V_T value than for a completely closed stand of trees. However, this hypothesis still needs to be tested experimentally.

2.3.4
Leaf Area Measurements

Several instruments measure the leaf area of a stand, but measuring the leaf area of an isolated tree is more difficult. Empirically correlating leaf area to trunk diameter for a particular species offers an easy approach to making this measurement. Total tree leaf area is related to the number of active xylem vessels supplying water to those leaves. The leaf area-xylem relationship is morphologically evident in the size of the crown and the diameter of the trunk for a given species, but varies widely among species and tree conditions. For a ring-porous species, such as oak (*Quercus*), the amount of active xylem vessels is equivalent to the trunk diameter growth recent. For conifers and diffuse porous species, active xylem vessels may encompass the entire sapwood. An empirically derived relationship between total tree leaf area and trunk diameter would provide a useful equation that could be used to estimate leaf area by simply measuring trunk diameter. This leaf area trunk diameter relationship would likely be constant within a species, except under periods of prolonged stress that reduce trunk growth and leaf area.

2.3.5
Measuring Transpirational Water Use

Thermal dissipation probes are widely used to measure sap flow velocity through a cross-section of a tree trunk. The measurements are then used to quantify the rate of water use by the tree. Each probe consists of two needles containing chromel-constantan thermocouples. One needle contains a fine-wire resistance element heated with a constant voltage. The needles are inserted into holes drilled into the tree, with the heated needle closer to the top of the tree. Under high flow conditions the heat input into the upper needle is dissipated by sap flow, and the temperature difference (ΔT) between the upper and lower needle is reduced. Values for ΔT and sap velocity are related empirically [46][2]. The product of sap velocity and average trunk cross-sectional area between the two thermocouples yields the rate ($cm^3\ h^{-1}$) of sap flow.

The relationship between ΔT data obtained from TDPs and actual transpirational water use from containerized poplar trees has been investigated [26]. Preliminary results indicated that the empirical relationship [46] yielded sap flow values that were approximately 45% of the actual volumetric water use.

2.3.6
Demonstrating Groundwater Use

To some extent, trees may be opportunistic and use soil moisture in the vadose zone, especially after precipitation. One way to determine whether a tree is using groundwater or surface water is by measuring the molar ratios of the stable isotopes of hydrogen (deuterium to hydrogen), expressed as δD [47].

$$\delta D\ (‰) = [(R_{sample}/R_{standard}) - 1] \times 1000 \qquad (3)$$

Determination of the δD in the transpiration stream of woody plant tissues can be used to identify water sources where R_{sample} and $R_{standard}$ are the molar ratios of deuterium to hydrogen of the sample and standard, respectively. The standard is defined as Standard Mean Ocean Water (V-SMOW) and the units are parts per thousand (‰), commonly referred to as "per mil." The hydrogen and deuterium in plant xylem water is derived from soil water. Roots taking up water do not discriminate between the two isotopes [47].

In temperate climates, samples of rainwater and groundwater can be significantly different in isotopic composition. This difference arises because H_2O evaporates more rapidly than DHO [32, 34]. Therefore, surface water generally is more enriched in deuterium isotopes (per mil values are less negative) than groundwater. This isotopic ratio evidence has suggested that mature *Populus* and *Salix* rely exclusively upon groundwater even when water is available in the unsaturated zone. This phreatophytic pattern is observed in areas with groundwater as deep as 1 m bgs [32, 47].

[2] A dimensionless flow index, K, is calculated for a specific time period according to the relationship: $K = \frac{\Delta T_{Max} - \Delta T}{\Delta}$, where ΔT_{Max} is the value in a given time period (e.g. the pre-dawn hours) where ΔT is the greatest (zero set point). A value for sap velocity, V ($cm\ s^{-1}$), is then estimated using the following empirical relationship developed by Granier [46]: $V = 0.0119 \times K^{1.231}$. Sap velocity is multiplied by the cross-sectional area of the tree (basal area) at the location of the probe (As) to yield Fs ($cm^3\ h^{-1}$), the rate of sap flow, using the following equation: $Fs = As \times V \times 3600\ s\ h^{-1}$. Thus, the parameter, Fs, represents the volume of water transpired per unit time.

3
Calculating and Modeling Hydraulic Control

One method to evaluate the effectiveness of a groundwater phytotechnology system is to use the estimated or measured water use rates (see Section 2.3) to estimate the groundwater capture zone dimensions using mathematical calculations or computer model simulations. A distinction should be made, however between capture-zone "analysis" and capture-zone demonstration. Where practicable, it is best to obtain actual field groundwater quality and/or hydrogeologic data to confirm the effectiveness of a containment system.

3.1
Mathematical Calculations

Groundwater flow and capture zone calculations can be used to evaluate the ability of a tree phytotechnology system to capture and contain a groundwater plume.

3.1.1
Darcy Flow Calculations

The average, steady state groundwater flow within the plume can be calculated using a form of Darcy's Law [48]:

$$Q_a = K\,I\,A, \tag{4}$$

In Eq. (4), Q_a is the ambient, pre-stress groundwater flow ($L^3\,T^{-1}$), K is hydraulic conductivity ($L\,T^{-1}$), I is the hydraulic gradient (dimensionless), and A is the cross-sectional area of the plume, measured in the plane orthogonal to the flow direction (L^2). Due to the distortion of the flow field during pumping, the pumping rate required to contain groundwater at a certain location is approximately twice the rate calculated using Eq. (4) (see Sections 3.1.2 and 3.1.3).

The Darcy approach assumes that groundwater extraction from the capillary fringe and water table will create a vertical capture zone extending to the bottom of the plume. If the top of the plume is below the water table, then the vertical thickness used in the calculation must include the entire vertical distance from the water table down to the bottom of the plume, including the "clean" zone above the plume.

3.1.2
Two-Dimensional Capture Zone Calculations

For the purpose of evaluating phytotechnology alternatives, two-dimensional (2-D) capture zone solutions should be used only for thin aquifers in which the vertical flow component may be neglected. In this case, a stand of trees can produce a 2-D capture zone similar to a fully penetrating pumping well.

Several researchers have presented analytical solutions for the steady state, two-dimensional capture zone affected by a fully-penetrating pumping well in a

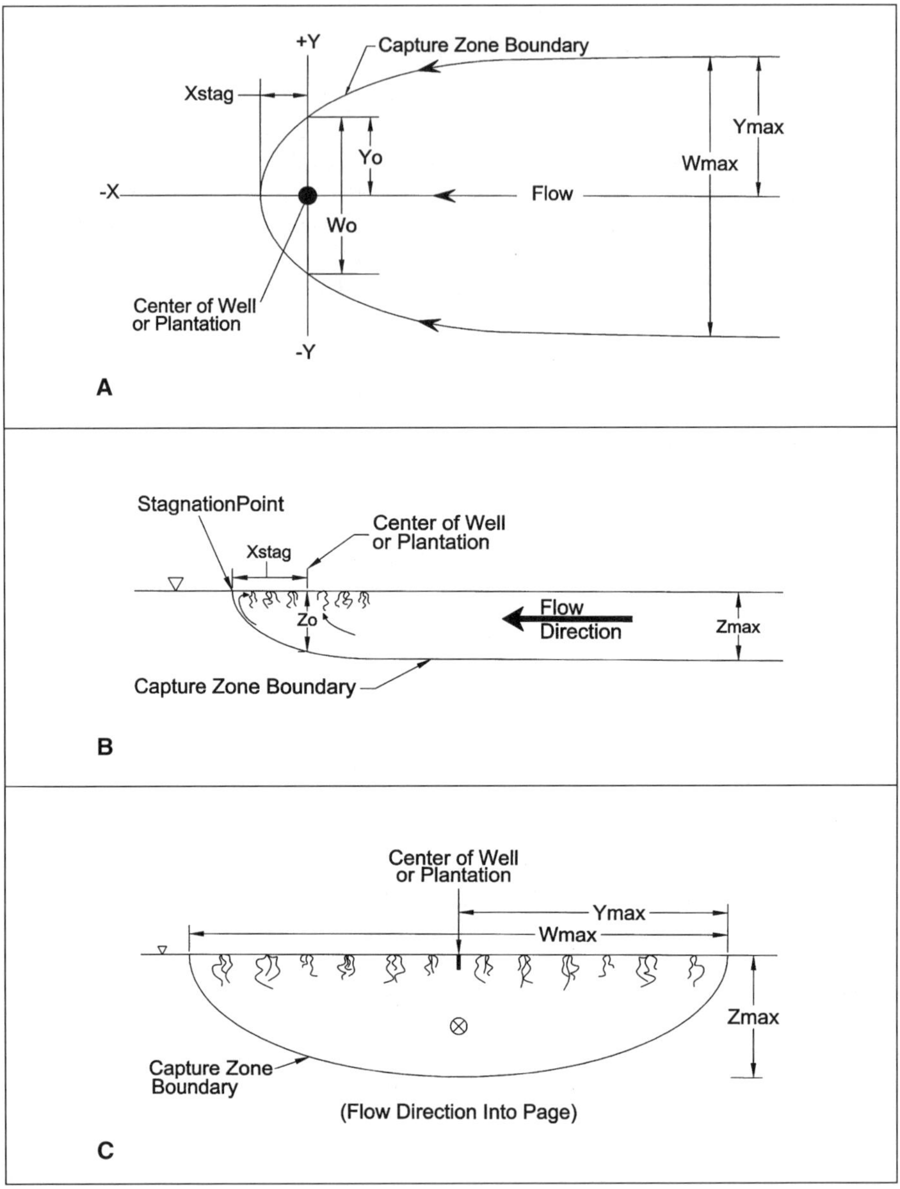

Fig. 2. Steady State 3-D Capture Zone Created by a Partially Penetrating Pumping Well or Transpirational Water Use by a Tree Stand (**A** = plan view; **B** = cross section parallel to flow where the top of the parabola is cut off at the water table or upper confining layer; **C** = cross section perpendicular to flow with width W_{max} and depth Z_{max})

confined aquifer [49, 50]. These methods assume uniformly horizontal groundwater flow. Extracting groundwater with horizontal flow results in a capture zone that is parabolic in plan view (see Fig. 2A).

The geometric elements of 2-D capture zones are summarized in Table 1. The steady-state capture zone extends to the bottom of the aquifer, and the width reaches a maximum value (W_{max}) upgradient from the extraction location. The groundwater flow through the cross section, defined by W_{max}, is equal to the groundwater extraction rate. The width of the capture zone measured cross-gradient from the extraction location (W_0) is one-half of W_{max}. Thus, containing a certain width of the aquifer at the extraction location requires pumping twice the rate of flow through the zone that exists under non-pumping conditions. The distance from the pumping center to the stagnation point (X_{stag}) defines the downgradient extent of the capture zone.

3.1.3
Three-Dimensional Capture Zone Calculations

Vertical flow in a thick aquifer can influence capture zone geometry, necessitating the use of a three-dimensional (3-D) groundwater flow analysis. The three-dimensional analytical methods assume horizontal flow prior to pumping. Three-dimensional capture zones of tree stands can be approximated using an analytical solution for a pumping well that penetrates a small fraction of the saturated aquifer thickness. The steady state, 3-D capture zone of a partially penetrating well screened at the top of the aquifer is parabolic in plan view (see Fig. 2A). The capture zone attains a maximum depth (Z_{max}) and width (W_{max}) at a location far upgradient of the pumping well or plantation (see Fig. 2B). The groundwater flow through the half-ellipse, defined by Z_{max} and W_{max}, is equal to the groundwater extraction rate (see Fig. 2C). A similar cross section exists perpendicular to flow at the pumping well location, with width W_0 and depth Z_0. The capture zone extends downgradient by the distance X_{stag}. While the calculations discussed in this section assume horizontal flow, Z_{max} would be larger in an area with an upward hydraulic gradient and smaller in an area with a downward gradient.

Three-dimensional capture zones of partially penetrating pumping wells in confined aquifers have been analyzed, including the case of a zero-penetration "point sink" situated at the top of an aquifer of infinite thickness [51]. Analytical solutions for anisotropic aquifers have also been developed in which the vertical hydraulic conductivity (K_z) is different from the radial, horizontal hydraulic conductivity (K_r). The capture zone elements of a zero-penetration point sink at the top of an aquifer are summarized in Table 1. The capture zone drawn through the pumping well location (semi-ellipse of width W_0 and depth Z_0) has an area of exactly one-half the area of the semi-ellipse defined by W_{max} and Z_{max}, further illustrating that containing a certain cross-sectional area of the aquifer at the pumping location requires pumping twice the rate of flow through the zone that occurs during non-pumping conditions.

Assuming an infinitely thick aquifer can lead to overestimating the vertical extent of capture and underestimating the lateral extent of the capture zone [51]. An analysis was presented of the 3-D capture zone of a partially penetrating well

Table 1. Methods for Calculating the Approximate Capture Zone Dimensions for a Single Pumping Well. These equations may be used to calculate the approximate 2-D or 3-D capture zone dimensions for a tree or stand of trees, as an analog for a pumping well. Capture zone dimension terms are illustrated in Fig. 3

Dimensions	2-D	2-D	3-D	3-D	3-D
Well Penetration	100%	100%	0%	10%	10%
Aquifer Type	Confined	Unconfined	Confined	Confined	Unconfined
Aquifer Thickness	Finite	Finite	Infinite	Finite	Finite
Condition	NA	NA	NA	$Q'a^{-1} \leq 2.5$	$Q'a^{-1} \leq 2.5$
X_{stag}	$-Q(6.28\,BU)^{-1}$	$-QL(3.14\,K_r(h_2^2-h_1^2))^{-1}$	$-(Qa(6.28\,U)^{0.5}$	See Note 8	See Note 9
Z_{max}	B	B	$0.798(Q(Ua)^{-1})^{0.5}$	$2Q(3.14UY_{max})^{-1}$	See Note 9
Y_{max}	$Q(2\,BU)^{-1}$	$-QL(K_r(h_2^2-h_1^2))^{-1}$	$0.798(Qa(U)^{-1})^{0.5}$	$-2.125\,X_{stag}$	See Note 9
W_{max}	$Q(BU)^{-1}$	$-2\,QL(K_r(h_2^2-h_1^2))^{-1}$	$1.596(Qa(U)^{-1})^{0.5}$	$-4.25\,X_{stag}$	$2\,Y_{max}$
Z_0	B	B – drawdown	$0.707\,Z_{max}$	$0.735\,Z_{max}$	$0.610\,Z_{max}$
Y_0	$0.5\,Y_{max}$	$0.5\,Y_{max}$	$0.707\,Y_{max}$	$0.680\,Y_{max}$	$0.633\,Y_{max}$
W_0	$0.5\,W_{max}$	$0.5\,W_{max}$	$0.707\,W_{max}$	$0.680\,W_{max}$	$0.633\,W_{max}$
$W_{max}(X_{stag})^{-1}$	6.28	6.28	4	4.25	4.38
Reference/Basis	49	67	51	51, 52	53, this study

1) NA = not applicable.
2) h_1, h_2 = pre-pumping unconfined aquifer thickness at two piezometers, upgradient and downgradient, respectively, from pumping well (both situated along a groundwater flow-line; base of aquifer assumed horizontal).
3) L = distance between two piezometers where h_1 and h_2 measured.
4) B = pre-pumping aquifer thickness at pumping well.
5) $U = K_r I$, where K_r is the radial (horizontal) hydraulic conductivity and I is the horizontal hydraulic gradient component.
6) $a = (K_r(K_z)^{-1})^{0.5}$, where K_z is the vertical hydraulic conductivity.
7) $Q' = Q(UB^2)^{-1}$.
8) Based on best-fit solution to data presented by Faybishenko, et al., (1995) [52], with extension to anisotropic, finite-thickness confined aquifers by Zlotnik (1997) [51].
$X_{stag} = -[0.0241(Q'(a)^{-1})^3 - 0.1774(Q'(a)^{-1})^2 + 0.5312(Q'(a)^{-1}) + 0.0257]\,B$.
9) Based on best-fit solutions to numerical model results presented by Bair and Lahm (1996) [53] for anisotropic, finite-thickness unconfined aquifers:
$X_{stag} = -[0.0006(Q'a^{0.5})^3 - 0.018(Q'a^{0.5})^2 + 0.3043\,Q'a^{0.5} + 0.0606]\,B$.
$Y_{max} = [0.0017(Q'a^{0.5})^3 - 0.0492(Q'a^{0.5})^2 + 0.3043\,Q'a^{0.5} + 0.1018]\,B$.
$Z_{max} = [0.0758(Q'a^{-1})^3 - 0.4427(Q'a^{-1})^2 + 0.9234\,Q'a^{-1} + 0.2486]\,B$.
$Y_0 = [-0.1278(Y_{max}B^{-1})^2 + 0.7728\,Y_{max}B^{-1}]\,B$, for $0.0\,B < Y_{max} \leq 1.5\,B$; otherwise $Y_0 = 0.633\,Y_{max}$.
$Z_0 = [0.2134(Z_{max}B^{-1})^2 + 0.5043\,Z_{max}B^{-1}]\,B$, for $0.0\,B < Z_{max} \leq 0.7\,B$; otherwise $Z_0 = 0.610\,Z_{max}$.
10) The 3-D capture zone solutions for finite-thickness aquifers presented in Table 1 are not recommended when $Q'a^{-1}$ is greater than approximately 2 for the confined aquifers or 2.5 unconfined aquifers. In these cases it may be more appropriate to use the fully penetrating well solution. Additionally, the equations summarized in the table are most applicable when the plan view footprint of the plantation is compact, such as square or circle. If the plantation consists of one or a few relatively long lines of trees aligned perpendicular to the flow direction, then the equations in Table 1 should not be used, and it would be more appropriate to use a cross-sectional (profile), 2-D analytical or numerical model, or a 3-D numerical flow model to assess the potential depth of capture beneath the trees.

in a confined aquifer of finite thickness for various degrees of well penetration [52]. By evaluating the capture zone elements in terms of the dimensionless flux, solutions were presented that are applicable to any aquifer thickness. The dimensionless flux (Q′) can be expressed as:

$$Q' = Q\,U^{-1}\,B^{-2} \qquad (5)$$

In Eq. (5), Q is the pumping rate (L^3/T), U equals Darcy velocity, K_r I, and B is the aquifer thickness.

These principles were extended to anisotropic, confined aquifers by modifying the dimensionless flux term as a function of the square root of the anisotropy, and the results are presented as a function of aquifer thickness (see Table 1) [51, 52]. This solution includes an expression for the stagnation point, or the furthest downgradient point along the capture zone boundary, in terms of $Q'a^{-1}$. The expression in Table 1 represents the "best fit" for the data presented for an anisotropic aquifer in accordance with observations [51, 52]. The capture zone is similar for any well penetration less than 20% of the aquifer thickness [52, 53].

The results of 3-D numerical simulations provide a basis for general capture zone solutions in unconfined aquifers (see Table 1) [53]. The results presented for a well penetrating 10% of the aquifer [53] were analyzed to obtain an "empirical" best-fit expression for each of the key capture zone geometrical elements based on a dimensionless flux term modified for anisotropy. Testing of these expressions confirms that a satisfactory reproduction of the numerical capture zone simulations [53] can be achieved using the empirical expression presented in Table 1.

3.2 Computer Model Simulations

While analytical calculations provide a means to estimate the geometric elements of a capture zone, computer simulations provide graphical output that can assist in visualizing a predicted capture zone.

3.2.1 *Analytic Element Method*

Analytic element method (AEM) models such as WinTran® [54] and TwoDan® [55] have graphical interfaces and provide a high degree of mathematical precision and flow-line resolution. AEM models solve for the head distribution within an aquifer assumed to have infinite lateral extent and can be used to simulate the potential hydraulic influence of one pumping well or a large group of individual extraction locations (i.e. "trees").

A relatively simple capture zone comparison was developed with WinTran®. The simulated 2-D capture zone predicted for one fully penetrating pumping well (see Fig. 3A) is compared with that of a tree stand (see Fig. 3B). The tree stand was simulated as 25 discrete, small diameter (3 mm), low-flow pumping wells spaced every 1.5 m. The capture zone widths of the single pumping well and

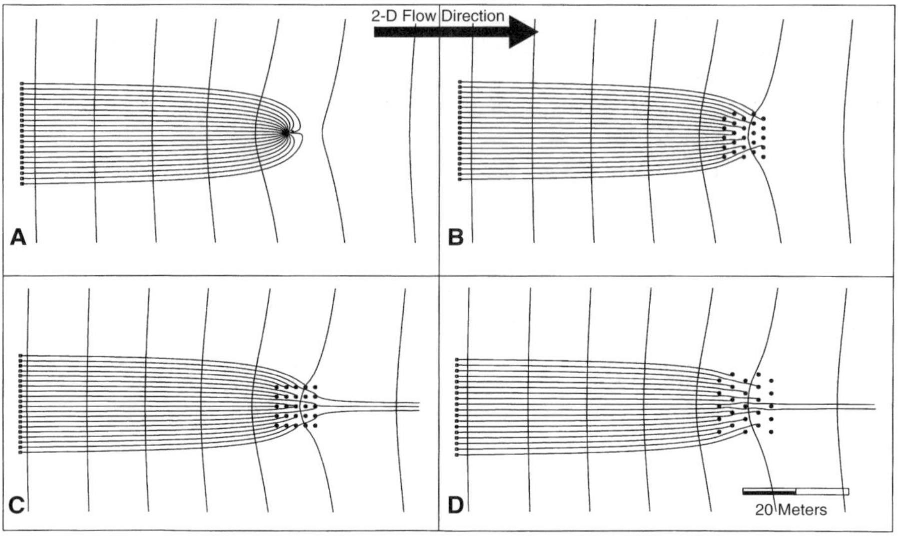

Fig. 3. Computer Simulations of Groundwater Flow Using WinTran® (comparisons of 2-D capture zones. **A** = one fully penetrating pumping well; **C** = a stand of 25 trees on staggered 1.5 m centers, simulated as low flow pumping wells; **B** = a stand of 25 trees in a grid pattern (1.5 m centers); **D** = a stand of 25 trees on staggered 2.25 m centers. Parameters: hydraulic conductivity = 1×10^{-4} m s^{-1}; hydraulic gradient = 0.005; thickness of saturated zone = 3 m; total pumping rate = 2800 L/d)

group of "trees" are similar. Figures 3C and 3D show that hydraulic gaps can occur when using excessive spacing between trees and a square grid rather than a staggered pattern.

Figure 4 shows a detailed depiction of hydraulic head contours within the tree stand simulated in Fig. 3B at a contour interval of 1.5 mm. Detailed particle tracking, using 100 equally spaced particles within the limits of the capture zone shown on Fig. 3B, confirmed that there were no hydraulic gaps within the simulated stand of trees. All 100 particles were captured. In spite of the complete capture of groundwater particles by the stand, the model predicted no measurable reversal of hydraulic gradient. Detailed inspection indicated a small zone with reversed gradient immediately downgradient of each simulated tree.

AEM models can also be used to simulate groundwater plumes and to evaluate their fate as they enter and move through a tree plantation. By running simulations at time intervals, it is possible to create a pseudo-transient, time-stepping effect to account for the change in extraction rate achieved by trees throughout the year. Figure 5 presents an example of solute transport. Hydraulic parameters used in these simulations are identical to those listed above for Fig. 3.

During the growing season, when the rate of transpiration is assumed to be in a steady state, the simulated plume is cut off by the phytotechnology system. In temperate regions, however, deciduous trees go dormant, transpiration stops, and the plume "escapes" during the winter months. This effect is presented in Fig. 5D,

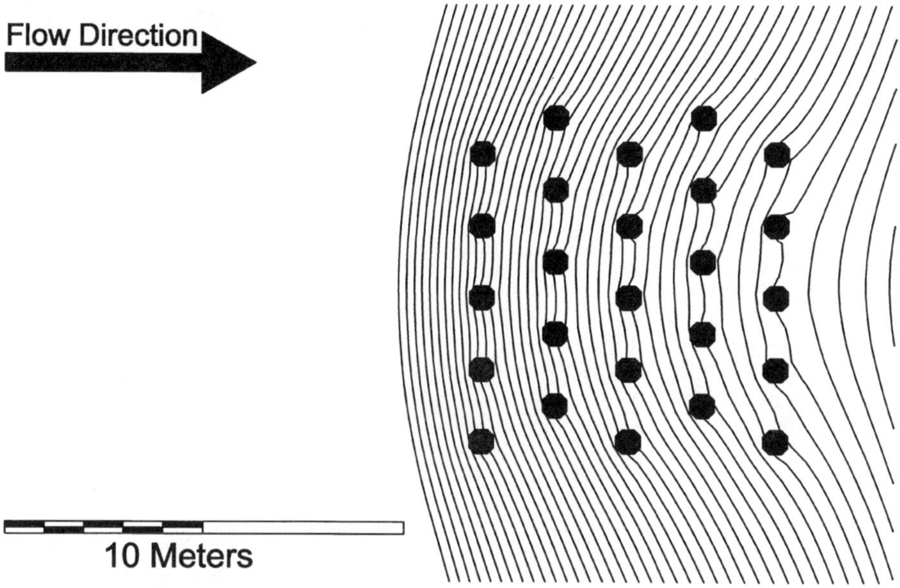

Fig. 4. Computer Simulation of Hydraulic Head (Elevation) Contours Using WinTran®. Detail of a stand of 25 trees on staggered 1.5 m centers (c.f. Fig. 3 B). Groundwater elevation contour intervals = 1.5 mm (the other hydraulic parameters are the same as for Fig. 3)

in which the plume resumed migration during a 6-month dormant season and traveled beyond the downgradient edge of the tree stand. However, by extending the footprint of the tree stand in the direction parallel to groundwater flow, as shown on Fig. 5E, the problem of dormant-season plume advance is addressed. With the extended plantation, by the end of the growing season, the plume boundary is back to where it started at the beginning of the previous dormant season.

These simple 2-D solute transport simulations have important implications for designing phytotechnology remedies in temperate regions where deciduous trees have a prolonged dormant season. Phytotechnology designs must consider the seasonal oscillations of plume growth and ensure that the length of the stand is sufficient to consume the plume that is extended during the dormant season. The optimal length of a tree stand is therefore site-specific and involves estimating solute velocity and travel distance during the dormant season.

Consumption of groundwater by phreatophytic trees varies on a diurnal basis (Section 2.2.1). Drawdown takes place during the day as a result of phreatophytic consumption, and recovery takes place at night when stomata are closed. However, these fluctuations can be neglected in evaluating and designing groundwater capture-zones by plantations of trees, provided one uses the average daily water removal rate for design purposes.

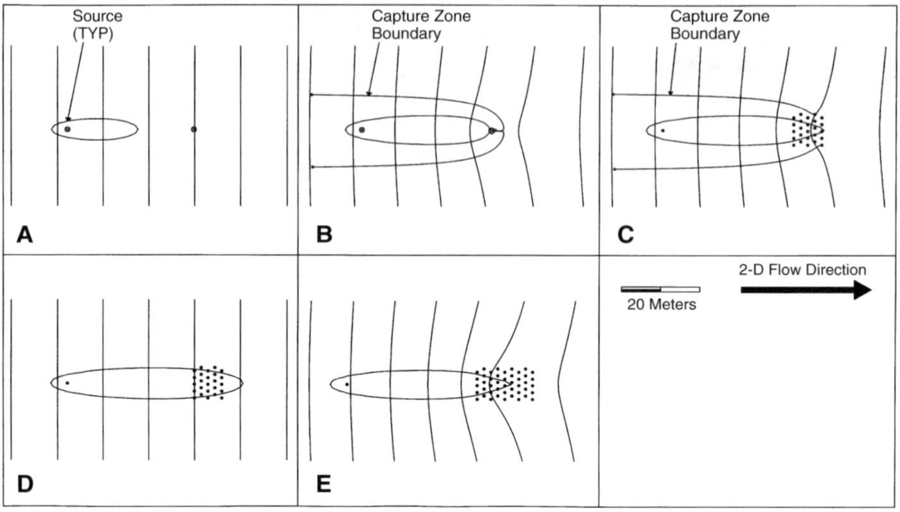

Fig. 5. Computer Simulation of Groundwater Plume Containment Using WinTran® (**A** = dissolved-phase groundwater contaminant plume; **B** = hydraulic control of the plume simulated by a single pumping well; **C** = hydraulic control by a stand of 25 phreatophytic trees on staggered 1.5 m centers (c.f. Fig. 3 B); **D** = the plume "escapes" during a 6-month dormant season (i.e. plume travels beyond tree stand); **E** = the stand of trees is extended to 45 trees total). The solute in these simulations is a non-reactive, non-sorbing compound such as bromide. The longitudinal and transverse dispersivity values are 3 m and 0.3 m, respectively and the porosity is 0.3. The extended stand maintains long-term hydraulic control, and pulls the plume back by end of growing season

3.2.2
Numerical Models

Numerical models use a system of numerical "cells" to represent the geologic formation in two dimensions, and may include a series of stacked "layers" of cells to simulate flow in three dimensions. Hydraulic conductivity values can be assigned independently for all three coordinate directions. Three-dimensional numerical models can be used to assess the vertical extent of capture beneath a plantation installed above a relatively thick aquifer or in areas with a significant vertical flow component.

One of the most widely used numerical models is the United States Geological Survey's (USGS's) modular 3-D, finite-difference groundwater flow model [56], commonly referred to as MODFLOW. Numerical models such as MODFLOW offer less resolution of flow lines than AEM models in capture zone analysis. Numerical models represent the flow field as numerous interacting cells. Head values are computed only at the nodes, which are either at the center of the cell or at the intersection of the grid lines that define the cell boundaries. The MODFLOW heads file is typically input to a particle tracking code such as MODPATH [57] to visualize the flow field, and its accuracy depends on the discretization of the model grid as defined during the setup of MODFLOW.

Unlike AEM models that assume laterally infinite aquifer extent, numerical models require boundary conditions to be assigned. The boundaries at the edge of the model may be assigned as specified flux (including no-flow), specified head, or head-dependent flux. The upper boundary of a numerical model is commonly simulated using an assigned recharge term. Recharge is initially estimated and then used as a variable during calibration. The calibrated recharge rate accounts for the effects of precipitation, surface runoff associated with the soil type and slope, and transpiration due to vegetation. After the model is calibrated for key hydraulic parameters the effect of a phytotechnology system can be added to perform predictive simulations of groundwater flow and capture.

Options for simulating the influence of trees include the use of "well" (specified flux) cells, ET cells, or modified recharge. When using the well module in MODFLOW, the total expected water removal rate of the tree stand is specified at one cell or partitioned among an array of well cells. A negative, effective recharge term may be the best approach to simulating transpirational water use in numerical models because it is automatically applied to the highest active model layer. Recharge is negative when the sum of evapotranspiration and runoff exceed precipitation.

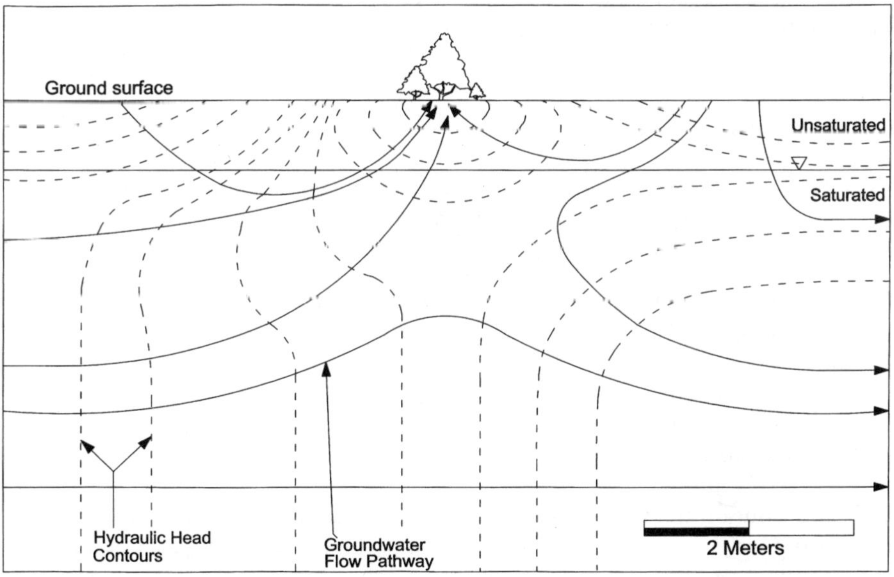

Fig. 6. Computer Simulation of 2-D Hydraulic Gradients Using the VS2DT Model (transpirational water use by a tree stand planted on sandy loam soil created a steady state capture zone that extends well beneath the water table)

3.2.3
Saturated/Unsaturated Flow Simulations

The USGS's VS2DT model is a variably saturated 2-D computer code that simulates solute transport in a variety of soil types in the presence of transpiration. VS2DT can simulate tree stand transpiration using a number of root parameters including root depth, root density, and root water potential. As shown in Fig. 6, this tool computes hydraulic gradients in both the saturated and unsaturated zones, and can be used to evaluate the thickness of the capture zones created by trees.

4
Demonstrating Hydraulic Control

Field measurements are a practical way to monitor and demonstrate the effects of a groundwater phytotechnology system. These potential hydraulic effects have been demonstrated in several ongoing field projects that have recently been installed at various sites, primarily in the United States.

4.1
Field Measurements

Although estimations of hydraulic control (Section 3) are a good starting point for determining hydraulic control, field measurements generally produce a greater degree of accuracy and should be used to determine the extent of groundwater capture whenever possible.

4.1.1
Plume Monitoring

Periodically sampling groundwater at monitoring wells located near the fringe of the plume indicates if the extent of the plume is increasing, decreasing, or remaining the same. The sampling protocol should be sufficient to:

- Compare data from similar seasons obtained for several years.
- Obtain samples in spring to determine the maximum annual extent of the plume.

4.1.2
Hydraulic Head Measurements

Hydraulic containment can be evaluated by measuring hydraulic head values at monitoring wells immediately downgradient from the pumping well or stand of trees. These data can indicate the location of the stagnation point, the farthest downgradient extent of the capture zone (see Fig. 2A). Between the pumping well (or center of plantation) and the stagnation point, the hydraulic gradient is reversed, demonstrating that groundwater flow has been cut off locally due to groundwater extraction. In some cases, however, a plume may be contained even without a measurable reversal of hydraulic gradient, as shown in Fig. 4 and described in Section 3.2.1.

4.1.3
Tracer Tests

Another approach to demonstrate hydraulic control is a tracer test. A conservative, non-reactive solute such as bromide can be injected in an area immediately upgradient of a tree stand and then monitored at a line of wells bordering the downgradient edge of the plantation. Wells positioned within the stand can be used to monitor the progress of the tracer and to determine its extent of travel in the event that it does not exit the stand of trees. Furthermore, samples of plant tissue can be obtained to assess uptake of the tracer. In some cases, it may be practical to use a dye tracer, such as fluorescein, which would make it possible to use fluorescent monitors in the downgradient wells. This approach is similar to the plume monitoring approach described in Section 4.1.1, but entails creating and monitoring a tracer "plume."

4.2
Ongoing Field Projects

Several groundwater phytotechnology systems have been in operation for many years. These sites are being operated under the auspices of the US Environmental Protection Agency (US EPA) Superfund Program, the US Departments of Energy and Defense, private industry, and the consulting community. These examples, some of which are listed below, have formed the current level of understanding for groundwater phytotechnology systems.

4.2.1
US Army Aberdeen Proving Grounds (Aberdeen, MD)

A stand of 200 hybrid poplar trees was installed in 1996 in the flow-path of a contaminant plume with VOC concentrations ranging in the hundreds of parts per million [58, 59]. On-site weather conditions were recorded and sap flow measurements taken to determine rates of groundwater use by the trees. Researchers estimate that at maturity, the stand will transpire approximately 4162 L d^{-1} of groundwater, a pumping rate that is approximately one-fifth of that needed to achieve hydraulic control. However, modeling data and field evidence indicate that water use by the stand is seasonally containing the southeast edge of the plume [60].

4.2.2
Fuel Distribution Site (Ogden, UT)

A phytotechnology system was installed in Ogden in 1996 with the objective to control groundwater containing petroleum hydrocarbons [26]. Hybrid poplar trees were deeply planted in dense triple rows oriented perpendicular to the direction of groundwater flow. The stand (40 trees total) was never irrigated. Piezometers were installed to measure water table elevations and contaminant levels upgradient, within, and downgradient of the trees. In 1998, analysis of the root structure of a representative tree indicated that roots had extended down to the saturated zone, approximately 1.8 m bgs. The rate of water use by the stand during 1998 was estimated from ET_0, leaf area, and a water use multiplier (Section 2.3). Sap velocity data were collected to measure actual water use by the stand in late summer of 1998. Estimated values compared favorably with measured values, which averaged 10.5 L d^{-1} $tree^{-1}$. Net groundwater use by the stand in 1999 was estimated at 1684 L d^{-1} (41 L d^{-1} $tree^{-1}$). Although the trees transpired a volume of water equivalent to a 3 m thickness of the saturated zone, water table elevation data collected in 1999 did not indicate a depression in the water table.

4.2.3
US Geological Survey Site (Charleston, SC)

Approximately 600 poplar whips were planted over 0.16 ha in December 1998 to help control groundwater contaminated with total petroleum hydrocarbons (TPH) and BTEX compounds (benzene, toluene, ethylbenzene and xylenes) [61]. The groundwater at the site is 0.6 to 1.2 m bgs. Based upon calculations using the Penman equation and weather station data, leaf temperature, and leaf wetness data, researchers estimated the 1-year-old trees removed 3785 L d^{-1} of soil moisture and groundwater from the site in 1999. In contrast, a simple draw down model in conjunction with measurements of groundwater level fluctuation showed that the 1-year-old trees used only 189 L of groundwater d^{-1} Groundwater level fluctuation measurements were taken again in 2000, and using the draw down model, researchers estimated water use by the 2-year-old trees had increased to over 1135 L d^{-1} of groundwater [62].

4.2.4
Solvents Recovery Service of New England (SRSNE) US EPA Superfund Site (Southington, CT)

A groundwater phytoremediation system consisting of approximately 475 hybrid poplar trees *(Populus deltoides x P. nigra)* and 475 willow trees (*Salix alba*) was installed at the SRSNE site in Southington, CT in 1998 and 1999 [63]. Groundwater at the site contains a complex mixture of volatile organic compounds (VOCs), the majority of which are localized in a containment area upgradient of a sheet pile wall. Groundwater from the containment area is currently being pumped from a series of extraction wells to an ultraviolet (UV)/oxidation treatment facility. The purpose of the phytoremediation system is to decrease costs at the site by reducing the amount of water that needs to be treated by the UV/oxidation system and to facilitate biodegradation of the VOCs in the containment area. An inward hydraulic gradient can be maintained by pumping approximately 72 L min^{-1} from the containment area of 0.3 ha. By 2002 (the fifth growing sea-

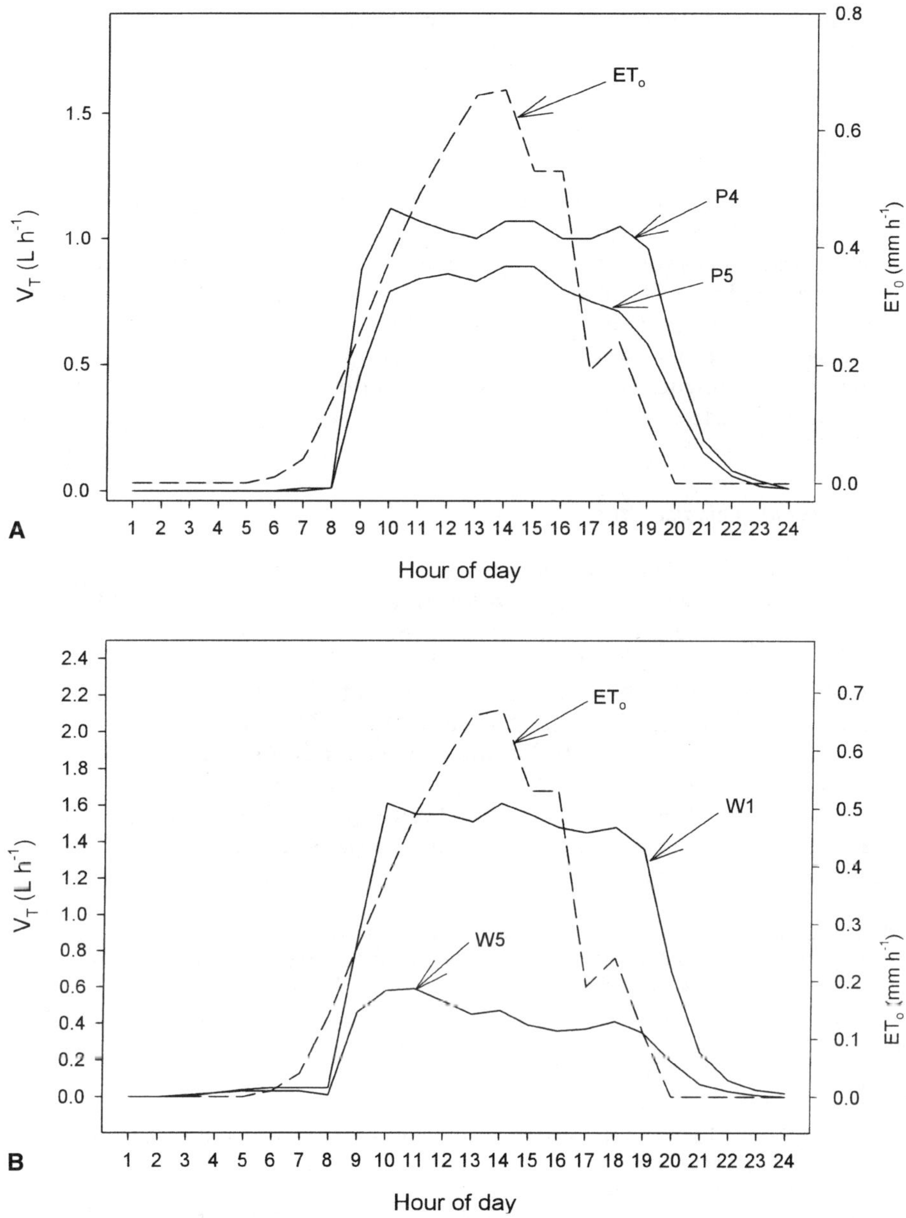

Fig. 7. Diurnal Cycles for Transpirational Water Use and ET_0 at the SRSNE Superfund Site (**A** = poplar trees; **B** = willow trees). The diurnal changes in V_T were compared for trees that generally had high (P4, W1) or low (P5, W5) apparent rates of transpiration, as measured by the TDPs. Values for V_T, averaged over the study period: P4, 22.9 L d^{-1}; P5, 9.0 L d^{-1}; W1, 36.1 L d^{-1}; W5, 10.5 L d^{-1}

son), the stand of trees is expected to use 34 L min^{-1} (average, April through October). Water use estimations for this site are shown in Fig. 7).

4.2.5
Argonne National Laboratory – East (Argonne, IL)

Over 800 hybrid poplars and willows were planted over a 1.2 ha area in 1999 to provide hydraulic containment of groundwater contaminated with VOCs and tritium. Groundwater modeling, using estimates of water use by the stand at maturity, indicated that year-round groundwater containment would be achieved and that de-watering of the aquifer could occur [64, 65]. Particle tracking in January 2000 showed that a large portion of the plume was being contained. Monitoring in the following years will determine if the stand will fully contain the plume or if more trees are needed.

4.2.6
MTBE / TBA Contaminated Site (CA)

The California EPA's State Water Resources Control Board collected data for a row of mature Monterey pine (*Pinus radiata*) trees that are growing downgradient from a leaking underground storage tank [66]. The concentrations of methyl-*tert*-butyl-ether (MTBE) and *tert*-butyl alcohol (TBA) in the groundwater reportedly decreased by three orders of magnitude as the contaminant plume passed beneath the trees, and detectable levels of MTBE and TBA were recovered in the transpiration condensate. Preliminary data indicated that the trees transpired 4 to 6 g of MTBE and TBA daily from the shallow groundwater. In addition, transpirational water use by the stand caused a measurable drawdown in the water table (the subsurface contains bay mud and silty clay and has relatively low hydraulic conductivity).

Acknowledgements. The authors would like to thank Janeece Christensen for her careful and thorough work in assembling and proofreading the manuscript, and Brandon Chard for his technical help with measuring water use parameters.

5
References

1. Waring RH, Schlesinger WH (1985) Forest Ecosystems: Concepts and Management, Academic Press, Orlando, FL
2. Yelenosky G (1964) Proc Int Shade Tree Conf 40:127
3. Kozlowski TT (1984) Extent, Causes, and Impacts of Flooding. In: Kozlowski TT (ed), Flooding and Plant Growth, Academic Press, New York, p 1
4. Yamamoto F, Kozlowski TT, Wolter KE (1987) Can J For Res 17:69
5. Alben AO (1958) Proc Am Soc Hortic Sci 72:219
6. Hall TF, Penfound WT, Hess AD (1946) J Tenn Acad Sci 21:18
7. Kramer PJ, Riley WS, Bannister TT (1952) Ecology 33:117
8. Opik H (1980) The Respiration of Higher Plants, Arnold, London
9. Chirkova TV (1968) Fiziol Rast (Moscow) 15:565
10. Armstrong W (1968) Physiol Plant 21:539

11. Philipson JJ, Coutts MP (1978) New Phytol 80:341
12. Kramer PJ (1983) Water Relations of Plants, Academic Press, New York
13. Zisa RP, Halverson HG, Stout BJ (1980) USDA For Serv Res Pap NE NE:451
14. Lyr H, Hoffman G (1967) Int Rev For Res 2:181
15. Canadell RQ, Jackson MB (1981) Alleviating Aeration Stresses. In: Modifying the Root Environment to Reduce Crop Stress, Am Soc Agric Engineers, St. Joseph, MI, p 141
16. Bowen B D (1985) Roots as a Component of Tree Productivity. In: Cannell MGR, Jackson JE (eds) Attributes of Trees as Crop Plants, Institute of Terrestrial Ecology, Huntingdon, England, p 303
17. Dobson MC, Moffat AJ (1995) Waste Mgmt Res 13:579
18. Fitter AH (1987) New Phytol 106:61
19. Nepstad DC, de Carvalho CR, Davidson EA, Jipp PH, Lefebvre PA, Negreiros GH, da Silva ED, Stone TA, Trumbore SE, Vieira S (1994) Nature 372:666
20. Stone EL, Kalisz PJ (1991) Forest Ecology and Management 46:59
21. Bibelriether H (1966) Allgemeine Forst und Zeitschrift 21:805.
22. Gregory PJ, McGowan M, Biscoe PV, Hunter B (1978) J Agric Sci 91:91
23. Canadell J, Jackson RB, Ehleringer JR, Mooney HA, Sala OE, Schulze ED (1996) Oecologia 108:583
24. Gasson PE, Cutler DF (1990) Arboricultural Journal 14:193
25. Lieffers VJ, Rothwell RL (1986a) Can J For Res 16:1201
26. Ferro AM, Chard J, Kjelgren R, Chard B, Turner D, Montaque T (2001) Internat J Phytoremed 3:87
27. Choudhury B, Monteith J (1986) Agric Meteor 36:215
28. Hatton T, Reece P, Taylor P, McEwan K (1998) Tree Physiol 18:529
29. Turner N, Schulze E, Gollan T (1984) Oecologia 63:338
30. Smith SD, Wellington AB, Nachlinger JL, Fox CA (1991) Ecolog Appl 1:89
31. White JWC, Cook ER, Lawrence JR, Broecker WS (1985) Geochim Cosmochim Acta 49:237
32. Busch DE, Ingraham NL, Smith SD (1992) Ecological Applications 2:450
33. Gregg J (1991) M.Sc. Thesis, University of Utah, Salt Lake City, UT
34. Dawson TE, Ehleringer JR (1991) Nature 350:335
35. Gatliff E (2000) Phytoremediation State of the Science Conference, Boston, MA
36. Tromble JM (1977) J Hydro 34:171
37. White WN (1932) U.S. Geol. Surv. Water Supply Paper 659
38. Trousdell KB, Hoover MD (1955) J For 53:493
39. Eastham J, Scott PR, Steckis R (1994) Agroforestry Systems 26:157
40. Bari MA, Shoefield NJ (1991) Agroforestry Systems 16:13
41. Hargreaves GL, Hargreaves GH, Riley JP (1985) J Irrigation Drainage Engineering ASCE 111:65
42. Monteith J, Unsworth M (1990) Principles of Environmental Physics, 2nd ed, Edward Arnold, London
43. Jarvis P, McNaughton K (1986) Adv Ecol Res 15:1
44. Caspari HW, Behboudian MH, Chalmers DJ, Renquist AR (1993) ASHS 118:562
45. Caldwell MM, Meister HP, Tenhunen JD, Lange OL (1986) Trees 1:25
46. Granier A (1985) Ann Sci For 42:193
47. Ehleringer JR, Dawson TE (1992) Plant Cell Environ 15:1073
48. Freeze RA, Cherry JA (1979) Groundwater Prentice-Hall, Englewood Cliffs, NJ
49. Keely JF, Tsang CF (1983) Ground Water 21:701
50. Javandel I, Tsang CF (1986) Ground Water 24:616
51. Zlotnick VA (1997) Ground Water 35:842
52. Faybishenko BA, Javandel I, Witherspoon PA (1995) Water Resources Research. 31:859
53. Bair ES, Lahm TD (1996) Ground Water 34:842
54. Environmental Simulations, Inc. (ESI) (1995) WinTran: Two-Dimensional Groundwater Flow and Contaminant Transport Model, Version 1.03. ESI. Herndon, VA
55. Fitts CR (1994) TWODAN: Two-Dimensional Analytic Groundwater Flow Model. Scarborough, ME

56. McDonald MG, Harbaugh AW (1988) Techniques of Water Resource Investigations. USGS, Book 6
57. Pollack DW (1989) USGS Open File Report, p 89
58. Tobia RJ, Compton HR (1997) Phytoremediation, International Business Communications, Inc., Southborough, MA p 91
59. Hirsh S, Compton H (2000) Phytoremediation State of the Science Conference, Boston, MA
60. Schneider WH, Wrobel JG, Hirsh SR, Compton HR, Haroski D (2000) Bioremediation and Phytoremediation of Chlorinated and Recalcitrant Compounds, Proceedings of the Second International Conference on Remediation of Chlorinated and Recalcitrant Compounds, Battelle Press, Columbus, OH, C2-4:477
61. Landmeyer JE (2000) Phytoremediation State of the Science Conference, Boston, MA
62. Landmeyer J (2001) USGS Pers. Comm.
63. Ferro AM, Chard B, Gefell MJ, Thompson BR, Kjelgren R (2000) Phytoremediation of Organic Solvents in Groundwater: Pilot Study at a Superfund Site. In: Wickramanayake GB, Gavaskar AR, Alleman BC, Magar VS (eds) Bioremediation and Phytoremediation of Chlorinated and Recalcitrant Compounds, Battelle Press, Columbus, OH, 2:461
64. Quinn J (2000) Phytoremediation State of the Science Conference, Boston, MA
65. Quinn JJ, Negri MC, Hinchman RR, Moos LP, Wozniak JB (2001) Internat. J. Phytoremed. 3:41
66. Parfitt DG, Arnold C, Kaltreider M (2000) Abstract from the Tenth Annual West Coast Conference on Contaminated Soils and Groundwater, San Diego, CA
67. Grubb S (1993) Ground Water 31:27

Received: December 2001

Vegetative Covers for Waste Containment

Steven A. Rock

United States Environmental Protection Agency – Office of Research and Development, National Risk Management Research Laboratory, 26 W. Martin Luther King Blvd., Cincinnati, Ohio 45268, USA. *E-mail: rock.steven@epa.gov*

Disposal of municipal and hazardous waste in the United States is primarily accomplished by containment in lined and capped landfills. Evapotranspiration cover systems offer an alternative to conventional landfill cap systems. These covers work on completely different principles than traditional covers do, and that difference may slow understanding and acceptance by site owners, regulators, and stakeholders. This chapter provides an introduction to this alternative technique and explains some of the common concerns regarding its implementation.

Keywords. ACAP, Alternative covers, Containment, Evapotranspiration, Landfill

Abbreviations

ACAP Alternative Covers Assessment Program
AEFC Alternative Earthen Final Covers
EPA Environmental Protection Agency
ET Evapotranspirative or Evapotranspiration
GCL Geosynthetic Clay Liner
K_{sat} Permeability at Saturation (cm/s)

Advances in Biochemical Engineering/
Biotechnology, Vol. 78
Managing Editor: T. Scheper

ORD	(US EPA) Office of Research and Development
P	Precipitation (mm/year)
PET	Potential Evapotranspiration (mm/year)
P_r	Percolation or Infiltration (mm/year)
R	Run-Off (mm/year)
RCRA	Resource Conservation and Recovery Act
RTDF	Remediation Technologies Development Forum

1 Introduction

Waste generation in the United States is about 4.3 pounds per capita per day. Despite efforts to encourage recycling, waste reduction, and other alternatives, landfill disposal is the primary method used to handle this large industrial, commercial, and municipal solid waste stream. There are uncounted former landfills, abandoned landfills, and dumps that were started, filled, and closed in a non-regulated or pre-regulation environment, leaving a legacy of potential environmental problems.

There are three main areas of regulatory concern regarding landfilled waste, and the same broad concerns apply to working, closed, or abandoned sites. These are most evident in the case of abandoned landfills. The main environmental hazard is the threat to groundwater. Precipitation either as rain or snow can infiltrate the surface of a waste mass, percolate through the mass, and pick up dissolved or suspended contaminants, and carry those contaminants to groundwater. This contaminated groundwater, or leachate, can travel to drinking water wells or emerge as surface water, sometimes carrying the pollutants for long distances. The second regulatory concern is the generation of gases through biological activity in a landfill. Organic waste such as food, yard waste, and paper can be consumed by microbes in the presence of water and in the absence of oxygen to produce methane and carbon dioxide, two primary greenhouse gases. Methane from landfills has also been known to cause explosions and fires when unintentionally concentrated and accidentally ignited. The third regulatory concern is physical contact with the waste, either through direct exposure to humans, by means of rodents and other disease vectors, or through litter scattered by wind. As the US Environmental Protection Agency (EPA) is charged with protecting human health and the environment, regulatory strategies have been to prevent or mitigate these three concerns.

In 1976, the US Resource Conservation and Recovery Act (RCRA) established regulations for landfill establishment, operation, and closure. The protective strategy adopted by the RCRA was to surround the waste with an impermeable water and gas tight wrapping that prevented aqueous and gaseous transport pathways. This effectively isolated the waste from water which triggers both gas production and leachate formation. This was to be accomplished through a system of bottom liners, careful layering of waste, engineered slopes, and an impermeable cover. Any precipitation that landed on the site would be diverted away from the waste, leaving the contained material dry, isolated from transport mechanisms, and biologically inactive. There are many variations of materials and techniques that can be employed to accomplish this isolation, all of which may be grouped as conventional covers.

In a conventional cover system, the water balance for a site is very simple: the input of water in, in the form of rain or snow (precipitation) = equals the outflow of water out in the form of run-off. There is an alternative cover paradigm that may be as effective in protecting groundwater, isolating waste from contact, and preventing gas generation. There may, in fact, be several effective alternate paradigms, although this chapter only focuses on the evapotranspirative, or (ET) cover. Also known as vegetative covers, ET covers work by engineering the landfill cover into a site-specific water handling system utilizing a vegetated ecosystem.

2 Alternative Cover Paradigm

Using planted technologies to effect environmental improvement is not startlingly new. People have been growing plants to modify their environment for millennia. In contrast to a conventional cover, The ET cover system does not aim for total exclusion of water from the site, in contrast to a conventional cover. The system, which is sometimes called a 'sponge and pump' in contrast to the conventional cover 'raincoat' cover, allows a certain amount of water to be stored in the soil-root layer or rhizosphere, where it is held until the plants can use and transpire it. The water balance for an ET cover is somewhat more complicated than that of the conventional cover: the input of water equals the interception by plants, plus run-off, plus storage followed by evapotranspiration. Evapotranspiration is the combination of evaporation that would occur in a particular spot in the absence of plants, and the transpiration that occurs as plants process water for nutrient transport, cooling, and structure. The rate of ET depends on plants species and placement, as well as the climatological characteristics of temperature, wind speed, humidity, and growing season. Evapotranspiration can be estimated using the Pennman-Monteith equations.

Because ET covers will work differently in different areas, all installations need site specific designs. Climate-based applications can be broadly understood on a regional basis, in this case by generally separating a region (parts of the country or continent) by annual precipitation or by native climax ecosystem (see Fig. 1). The best candidate sites are those where the evapotranspiration from the native climax vegetation exceeds precipitation. Generally speaking, those parts of the country region that receive between 10 and 20 inches of precipitation each year are considered semi-arid with a native climax ecosystem of prairie grassland. Areas with more than 20 inches per year are classified as humid, with a native climax of woodlands or forest. Less than 10 inches per year is possibly too arid to sustain vegetative cover, although evaporation in very arid areas tends to exceed precipitation by far.

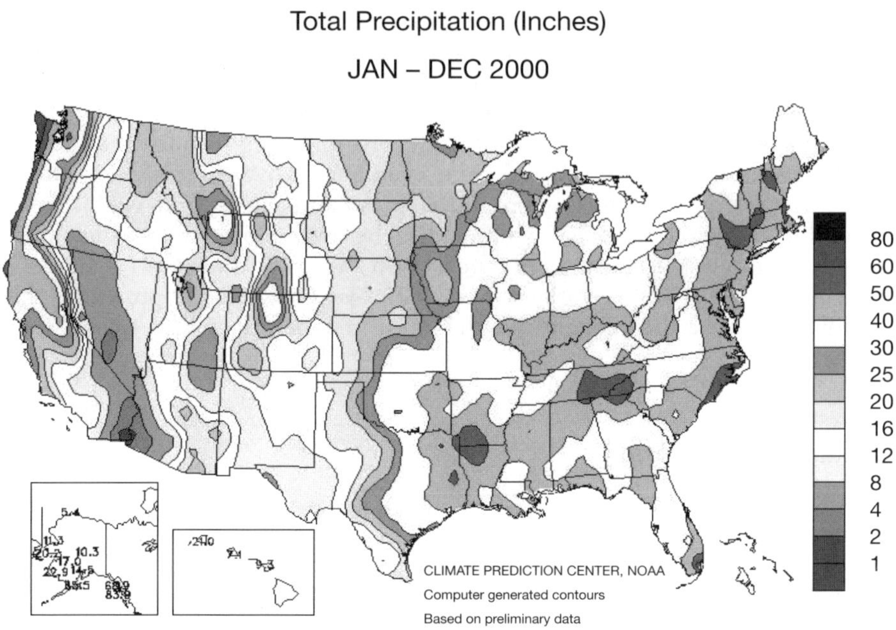

Fig. 1. US Precipitation Patterns [1]

3 Site Specific Design

Design of an ET cover in any climate depends on the water holding capacity of the soil. Water must be stored in that layer of soil that is in the range of the influence of the plants, either in direct contact with the roots, or within range of the capillary suction that plants can exert. Plant root influence also depends on the soil characteristics. For example, a sandy soil will allow easier penetration by roots, while a more silty soil will hold more water but restrict root penetration. Climate determines the types of plants available, while soil type determines the water holding capacity and hence, the depth of the soil cap needed to store the water. Therefore, the depth of the soil layer influences the type of plants chosen by dictating the necessary root architecture.

In arid and semi-arid prairie grasslands, there can be a tremendous variety of plants present. Some thrive during the cool and wet spring months but then yield to hot weather species during the summer. Some spread quickly into disturbed areas while others wait for the shade provided by the early species. Some have wide shallow root systems and some extend long roots that give them the capacity to withstand droughts. These plant characteristics help determine which species or combination of species will do the job of tapping into and using the water that will be stored in the cover system. Prairie species, like most plants, have most of their roots in the top three feet of soil, although some grasses send roots 30 feet or deeper. In semi-arid areas, most ET covers are designed with a three to four foot water holding layer , and that is vegetated with a variety of plant species. In wetter climates, a thicker soil layer is needed to capture the greater precipitation. Since trees have a greater root structure than grasses to support their larger biomass, designers of ET covers can use more depth for water holding. It is possible to design a water storage layer of up to eight feet thick that may be within the root zone of some trees.

A storage component of an ET system that relies on a very thick soil thickness layer for water holding capacity has an additional consideration. It may well be uneconomical to install such a layer. One type of design uses an installed soil layer that is two to four feet thick. In addition, this design relies on the existing cover and the waste mass itself to provide additional water storage capacity. In some older landfills that were closed with a soil cap before RCRA laws were enacted with a soil cap or simply abandoned without proper closure, the boundary between the surface soil and the waste is can be indistinct. On several sites that have clean soil containing increasing percentages of waste with depth, the US EPA has noted trees growing quite well, presumably sending roots into and amongst the waste. Among these sites are the Center Hill landfill in Cincinnati, Ohio that received municipal waste and incinerator ash; the Ohio EPA lead (Green II) site called Green II near Logan, Ohio that received municipal and hazardous wastes, and the US federal Superfund sites known as the Industrial Excess Landfill and the Woodlawn Landfill, both having received mixed industrial and municipal waste. Each of these sites did not discourage incidental vegetation and supported levels of grasses and trees that apparently rooted into the waste layer without detrimental effect to the plants.

The most common trees proposed for use on ET covers are hybrid poplars or hybrid willows. These trees are members of the Salicaceae family. They are hydrophilic and phreatophyitic which means that they tend to thrive in water rich areas, are undamaged by overwatering or inundation, and they withstand drought with a deep and extensive root architecture. Despite popular misconceptions, tree roots do not seek water, nor do they have a way of sensing water behind barriers. Trees can and will follow water such that when a slug of precipitation is descending through a soil column, roots will follow and extend as deep as necessary to obtain sufficient moisture. Some groundwater fluctuates annually or seasonally. Therefore, a rising water table may inundate tree roots. Many tree species will shed or slough off roots that are under water and deprived of oxygen. Furthermore, some non-Salicaceae trees may even be killed under these circumstances. Phyreatophytes maintain their roots even when saturated and are still in place when the water table descends. Therefore, they are immediately ready to draw water from the deeper seasonal water level. No trees, not even phyreatophytes, will extend their root systems farther than necessary to obtain water. Similarly, trees will not penetrate into a saturated zone if their water needs can be met in the vadose zone.

Methods for planting a cover that rely on vegetation must be site specific like the design. The methods must be chosen for the local microclimate, time of the year, soil characteristics, and topography. Planting techniques and timing depend on the site and the species selected. For trees of the Salicaceae family, there are a number of ways to plant these species successfully, and at least an equal number of ways to be unsuccessful. Trees may be purchased in a wide variety of sizes from 18 inches to 12 feet tall and are available from a number of commercial and public growers. These species may be planted two feet deep or deeper, depending on the type of excavation devices used. For many years, the pulp, paper, and lumber industries, as well as the US Forest Service have researched cultivation methods for short rotation woody crops. It has been estimated that thousands of hybrid poplar varieties that have been groomed through natural genetic selection in order to thrive in a wide range of conditions. Because these trees root from cuttings and grow quickly (eight feet a year is not uncommon), they are useful for many applications such as wind breaks and privacy screens, as well as for wood products. The long experience of the wood industries in developing and using wood products and that of the US Forest Service means that landfill cover installers have a number of resources from which to obtain advice, instruction, and plant material.

The design of the cover system must also take into account the seasonal nature of planted systems. Depending on the climatological location, the "pumping" or active transpiration phase of the growing season may be between five and seven months, when ET greatly exceeds precipitation. During the remainder of the year, any precipitation must either be run-off, evaporated directly from the soil, or be stored in the soil. This seasonal storage must have the capacity to last throughout the dormant season until the spring vegetation leafs out and the resumption of active growth during the spring. Some areas have precipitation patterns that allow for this to easily occur, such as in those parts of the Midwestern US where fall and winter weather is typically dry followed by Spring rains and Summer thunderstorms. Other parts of the country receive the bulk of their precipitation during the time that when plants have shut down. These climate locations may need too much storage capacity to make the system feasible.

4 Treatment vs Containment

Having plant roots in direct contact with the waste may raise regulatory and public concerns regarding the safety of the ET cover since the purpose of these systems is generally intended to isolate waste from contact. The general concern is that plant roots may take hazardous waste into their roots, transfer the chemicals to the above ground portions of the plant, then release a hazard into the environment. This concern can generally be addressed by carefully considering the characteristics of the plants, the subsurface environment, and of the chemical of concern. For example, root zones or rhizospheres are conducive to microbial activity. One result of a biologically active root zone is that methane, which may be generated by anaerobic decomposition in a landfill, can be consumed in the aerobic portion of the soil. Rates for methane generation and degradation are dependant on waste composition, soil moisture, gas transfer mechanisms through soil, and the number of methane-consuming microbes available.

Furthermore, few environmental contaminants chemicals are taken up in their parent form by plants, and even fewer are released intact. Some plants stimulate the biodegradation of chemicals in-situ while others transform contaminants within the plant tissues. Collectively, these processes are known as phytoremediation (see other chapters in this volume). There are some volatile solvents that have been measured in the transpiration stream from a plant, but the quantities have been extremely low, and the concentrations in the groundwater have been much higher than would be expected at most landfills. Some plants do accumulate metals in leaves and shoots which may be consumed by insects and other herbivores, but again, in a landfill situation, this is a very unlikely scenario as the rate and extent of accumulation is generally very small.

In a small number of cases, it may be the intention of the cover designer to include a treatment component as part of the plan. John Fletcher of the University of Oklahoma has recently proposed a cover system that would remediate waste as a "moving front" [2]. Basically, the system would clean the soil in a treatment phase, then use the clean layer for water storage capacity in a containment phase. Such a plan may be technically feasible, though awkward to regulate since the environmental rules typically tend to rigidly divide separate treatment from containment. One strategy that was successfully proposed and accepted by the Ohio EPA for a former landfarm was to demonstrate containment in part due to ET vegetation and allow phytoremediation to continue during the post-closure period. The expectation on the part of the land owner was that the planted "treatment cover" would not only contain the waste but, at some point, would remediate the contaminants to the target clean-up levels determined by the state. The site owner could then close the site under the clean closure rules and thereby discontinuing the maintenance of the cover and run-off collection system, as well as eliminating the periodic sampling events.

State regulations generally do not have provisions for these treatment covers. Similarly, there are no specific federal guidelines either for phytoremediation or for ET covers. Therefore, these technologies are typically only accepted and considered truly protective if they can show a performance that achieves the regulatory goals , that is, if the system is truly protective [3]. The US federal RCRA requirements, which are often the basis for state regulations, include a provision that landfill covers should follow a specific set of guidelines, "or equivalent" (RCRA 40 Code of Federal Regulations Part 258). There is no guidance as to how to demonstrate equivalence, or even what is the standard to which an alternative should be equivalent. The situation becomes more complicated when an alternative involves a completely different way to accomplish the same goal, as when an ET cover is compared to a conventional cover.

In an effort to gather data for eventual guidance on alternative covers, in 1997, the US EPA Office of Research and Development (ORD) in conjunction with the Remediation Technology Development Forum (RTDF) launched the Alternative Cover Assessment Program (ACAP). Phase One of this multi-year study involved a survey of existing field testing systems sites in conjunction with an analytical and a comparison and analysis of existing computer models for predicting and evaluating the performance of various landfill cover systems. This survey discovered that despite 28 projects that measured alternative cover performance,

Fig. 2. ACAP Site Map

none of the results were nationally applicable, and few had any direct comparison to conventional covers. Similarly, while there were many computer models that have been used for alternative covers, none were consistently accurate for the unique situation of the ET cover. Most of the models tested did not account sufficiently for periodic, non-saturating water flows such as those following rain events. Furthermore, most did not incorporate the effects of vegetation, particularly, the daily and seasonal effects on water usage.

Phase Two of the project was the installation of twelve 12 field sites across the country during 1999 and 2000. These sites are located in eight US states which that span the range of climates from semi-arid to very humid, from southern Georgia to Utah to Oregon (see Fig. 2). Many of the sites have installed drainage lysimeters that can directly compare two cover systems such as an ET cover constructed immediately adjacent next to a conventional cover. Phase Three involves collecting and analyzing the data from the test facilities. More information on ACAP can be found on the internet at www.DRI.edu or www.RTDF.org.

Each site involved in the ACAP has a site specific design, usually paired with an site appropriate conventional cover for the site. Since the climatic range of the different test sites is broad, there is an equally great variety in ET cover configurations. Figure 3 shows some cover designs that ACAP is currently evaluating.

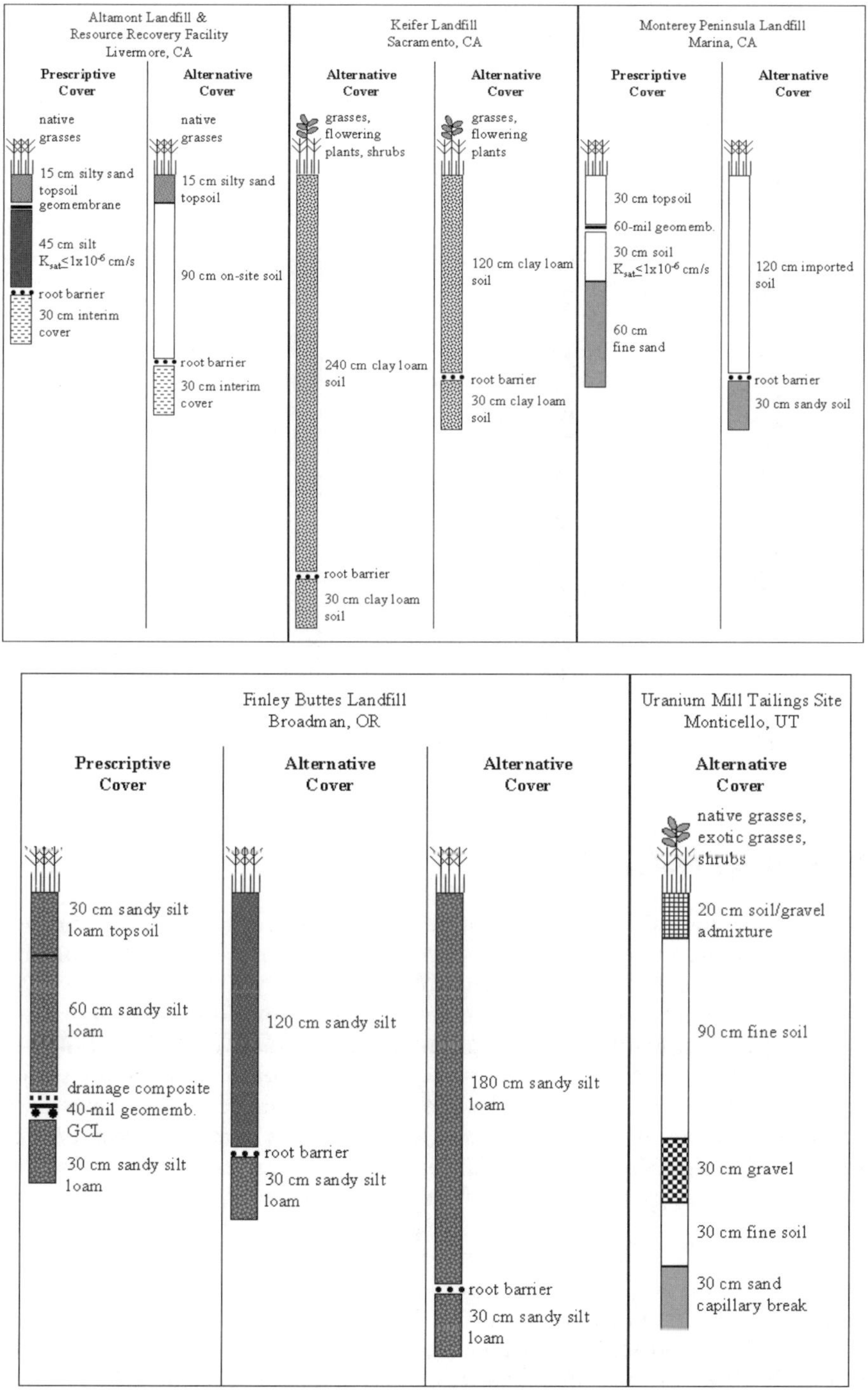

Fig. 3. Example evapotranspiration cover profiles

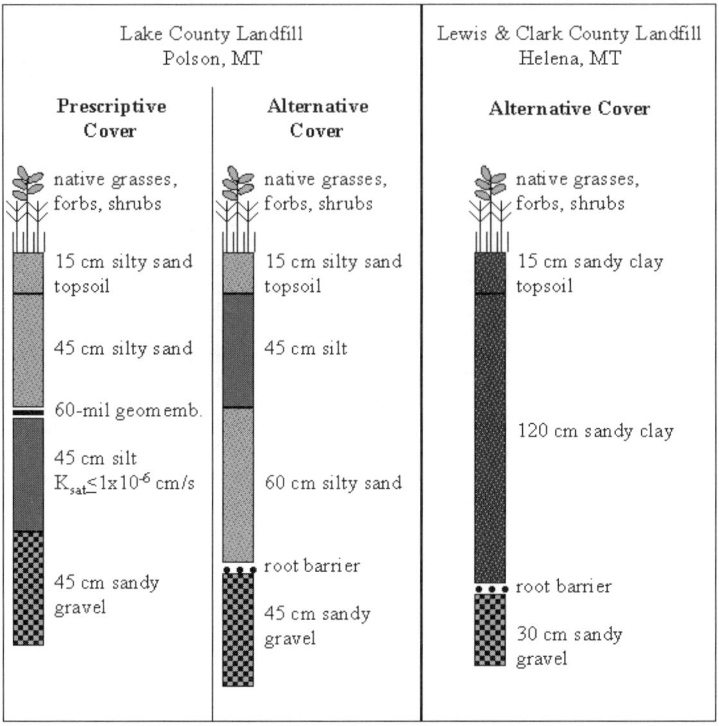

Fig. 3 (continued)

5
Water Balance Evaluation

Although percolation from a prescriptive cover is site specific, some estimates on typical percolation rates that can be expected can be derived from literature and calculation for the US Environmental Protection Agency's Alternative Cover Assessment Program (ACAP). ACAP operates on the assumption that an alternative cover is equivalent to a soil cover (e.g., an impermeable cover design) if the percolation rate is less than 10 mm/year in semi-arid and drier climates or less than 30 mm/year in humid climates [4–5]. In order to estimate percolation rates, a water balance evaluation can be conducted which consists of measuring and/or estimating all variables in the water balance including precipitation (P), run-off (R), evapotranspiration (ET), and percolation or infiltration (P_r).

$$P = P_r + ET + R \quad (1)$$

Each of the factors is variable in accuracy depending on the installation, location, and quality of instruments, and time for monitoring and analysis.

The precision and accuracy of precipitation (P) measurements depend on the location at which the measurements are made, the method used to make the measurements, the form of precipitation (rain vs snow), and the duration and frequency of the precipitation event. Inadequate wind shielding of rain gauges can also bias the precipitation measurement by 50% or more. With appropriate shielding, errors as large as 30% can still exist at high wind velocities (greater than 8 m/s). Even under ideal conditions, precipitation measurements have a precision less than ±10% [6].

Methods of estimating evapotranspiration (ET) vary in precision and accuracy as well. Potential evapotranspiration (PET) can be estimated with reasonable accuracy given accurate measurements of wind, solar incidence, humidity, and precipitation. However, this is not entirely applicable since at most sites ET will be less than PET during at least some portion of the year. There are commercially available devices and methods for estimating water use from crop plants, grasses, and trees, which may be employed to derive a reasonable site-specific estimate of ET at a given time and place. Extrapolation from point data to a season or a year is best done conservatively. For example, sap flow measurements from a stand of cottonwood trees at a USAF/USEPA SITE project in Texas indicated that water usage by those trees varied considerably from day to day and month to month depending on weather and water availability. During the mid summer months of the first full growing season, each tree used an average of 3.9 gallons per day, while in October, the same trees only used 0.24 gallons per day per tree [7]. Only after measuring for several entire days each month during two growing seasons can was there be any confidence in predicting the water usage of those trees over time.

Surface run-off (R) can be measured if using a catchment system such as is built using diversion structures that prevent run-on from adjacent areas and capture run-off for measurement. There are standard engineering techniques for diverting and measuring run-off with a high degree of accuracy.

Finally, percolation (P_r) can only be measured by catchment in the subsurface, which requires the construction of a test facility or lysimeter designed for that purpose. Alternatively, the water balance equation can be arranged to solve for P_r. By measuring or estimating the other components, a prediction of percolation can be achieved. Craig Benson of the University of Wisconsin offers the following example in his 2001 Article Field Evaluation of Earthen Final Covers [6].

> This provides a basis for estimating the precision with which percolation rates can be determined using the water balance method. Consider an AEFC alternative cover one meter thick at a humid location with 1200 mm/year of precipitation and a similar AEFC cover at a semi-arid site with 200 mm/year of precipitation. Assume that run-off and storage are measured with the precision attainable with current technologies (both 2%), ET is estimated with a constitutive method employing measured properties of the soil and vegetation (precision ~5 mm/year), and precipitation is measured with a quality shielded rain gauge with an error of 10%. Under these assumptions, the water balance method can be used to estimate percolation with a precision of 170 mm/year at the humid site and 50 mm/year at the semi-arid site. When the error in precipitation is ignored, the precision is 50 mm/year for the humid site and 30 mm/year at the semi-arid site.

However, these estimates of precision need to be viewed in the context of the performance criteria stipulated for Alternative Earthen Final Covers (AEFCs). In most cases, AEFCs used in place of a prescriptive cover are required to be "equivalent" to the prescriptive cover. Specifically; that is, percolation from the AEFC must be less than that from the prescriptive cover. Most importantly, all of these equivalency criteria are smaller than the precision with which percolation can be measured with the water balance method. Thus, most performance assessments made using the water balance method have insufficient precision to demonstrate that the required performance of an AEFC is being achieved.

6 Operation and Maintenance

Conventional covers do require the establishment and care of a vegetative cover, typically a grass cover on a foot of soil. However, planted systems have inherently different maintenance requirements from these conventional engineered covers. ET covers require somewhat more careful establishment of the vegetation, although there may be less ongoing maintenance once a self- perpetuating ecosystem is in place. Tree covers usually require 10–20% of the vegetation to be replanted following the first growing season. This percentage depends on climate conditions and whether irrigation is used.

Vegetated covers are susceptible to damage caused by biological attacks such as leaf defoliation from insects, damage to young trees by voles, and consumption by beavers. Most of the damage done by animals would not be widespread nor permanent. Perennial plants consumed one year will generally re-grow the following year. Poplar and willow trees can re-grow from their stumps after being consumed or cut. The efficacy of transpiration can be compromised if there is a widespread loss of vegetation. Periodic inspections would alert an operator to an infestation or problem before it could impair cover the functionality of the cover. Vegetated covers may also be impaired by catastrophic events beyond the control of operators: such as disease, fires, and strong wind. Little can be done to prevent or predict such events beyond normal fire prevention precautions. Diseased plants can sometimes be removed to prevent spread, or insect carriers sprayed to prevent further infestation. Wind damaged or knocked over trees may need to be removed and replanted. Trees generally tend to not cause deep intrusions into the cover when they fall. Rather than pull up large portions of the root system, trees tend to either break above ground, or break shallow roots, leaving the cover more or less intact.

In addition to standard landfill duties, landfill operators will need to perform other operation and maintenance activities such as pruning and thinning. Since these are outside the normal activities conducted at landfill facilities, training and education into farming and/or forestry practices may be needed.

7
Additional Considerations

Various designs for using trees and other plants for ET Covers or contaminant treatment may be subject to patent restriction or licensing arrangements. These can include the following applications:

1 Capping landfills or waste disposal sites as a means of hydraulic control and leachate minimization.
2 Capping contaminated soils as a means of hydraulic control and minimizing contaminant mobility.
3 Stabilizing metals-contaminated soils.

ET cover applications have a fairly recent history with the first tree covers planted in the 1980s and the first prairie covers tested around the same time. Such a short field time means that designers, installers, and regulators alike have limited experience with these alternatives, and that there is limited research from which to make decisions. Although intuitively clear, it may be hard to make a convincing case based on existing data or models that an ET cover will work predictably well in a specific setting. Furthermore, any planted system will require time to develop. Grasses or trees planted in one year will not reach full capacity for several years. In some cases, time is not a critical factor while in others, a multi-year development time would be unacceptable. Some abandoned sites have developed sustainable ecosystems in the absence of human intervention. In a few cases it might be acceptable to shape the site minimally to encourage run-off and augment the existing vegetation.

The reason that alternatives are being considered to existing landfill covers are usually either cost considerations or technical impracticality of conventional covers. Although precise cost estimates are hard to find and difficult to trust, ET covers generally cost half or less to design, install, and maintain than appropriate traditional covers. Furthermore, long-term maintenance on landfill covers to prevent or repair damage from erosion and cracks or burrowing animals can be expensive. ET covers are somewhat self-repairing in that a thick vegetative cover discourages erosion and an unconsolidated monolayer of soil tends to fill in gaps that result from waste settling and earth shifting.

It is conceivable that a tree cover could be cultivated to produce commercial products. Wood could be harvested on a rotation that allows continuous landfill coverage, for example every year every fifth row of trees could be harvested. Cutting 20% of a tree cover may significantly decrease the system efficacy. Since poplar and willow trees can be harvested to encourage re-growth from their stumps, the cut growth vegetation would regenerate, thus ensuring a continuous supply of wood while maintaining the ET cover. Such a system could produce an additional income stream for landfill operators. Trees and woody shrubs also sequester carbon from atmospheric carbon dioxide. At some point there may be a commercial value in carbon credits for stored biomass. Another by-product of the establishment of an ET cover that mirrors a native ecosystem is habitat creation for birds, insects, and mammals. Diversity in plants translates into increased food and shelter opportunities for wildlife. Sometimes, public sentiment favors a wooded vista over the traditional closed landfill grassy knoll.

8 References

1. NOAA (2001) Regional Climate Maps: USA, Climate Prediction Center, National Oceanic and Atmospheric Administration, US Department of Commerce
2. Fletcher JS, Olson P, Leigh MB (1997) The Role of Phytoremediation in Intrinsic Bioremediation. In: Leeson A, Alleman BC (eds) Proceedings of the Fourth International In Situ and On-Site Bioremediation, Vol 4(2), Battelle Press, Columbus, OH, p 115
3. ITRC (2001) Phytotechnology Technical and Regulatory Guidance Document, The Interstate Technology and Regulatory Cooperation Work Group Document PHYTO-2
4. Bolen M, Benson C, Albright W, Gee G (1999) Phase I Report for the Alternative Cover Assessment Program, US EPA Contract No. 68-C5-0036, Science Applications International Corporation, Hackensack, NJ
5. Bolen M, Benson C, Albright W, Gee G (2001) Phase II Report for the Alternative Cover Assessment Program, US EPA Contract No. 68-C5-0036, Science Applications International Corporation, Hackensack, NJ
6. Benson C, Abichou T, Albright W, Gee G, Roseler A (2001) Internat J Phytorem 3(1):105
7. US EPA (2001) Phytoremediation of TCE at the Former Carswell Air Force Plant 4, Innovative Technology Evaluation Report (In draft)

Received: November 2001

Regulatory Evaluation and Acceptance Issues for Phytotechnology Projects

Felix W. Flechas[1] · Marisa Latady[2]

[1] United States Environmental Protection Agency, Region 8, Solid and Hazardous Waste Program, 999 18th Street, Denver, Colorado 80202, USA. *E-mail: flechas.felix@epa.gov*
[2] Wyoming Department of Environmental Quality, Solid and Hazardous Waste Division, 122 West 25th Street, Cheyenne, Wyoming 82002, USA. *E-mail: mlatad@state.wy.us*

The use of plant based systems in hazardous waste management or site clean up is growing at a rapid rate. Major hurdles to implementation will be encountered by proponents if essential regulatory policies and regulations are not understood or addressed early in the project development. Understanding the regulatory requirements or issues as they relate to innovative clean up approaches such as phytotechnologies are critical to the successful implementation and long-term monitoring of these technologies. This chapter will address the issues that regulators may have in applying phytotechnologies to site conditions or contaminants, designing phytoremediation projects, monitoring or maintaining the systems for the long-term, establishing performance criteria, and demonstrating achievement of performance goals.

Keywords. Phytoremediation, Phytotechnologies, Containment, Treatment, Point of compliance, Technical impracticability

Advances in Biochemical Engineering/
Biotechnology, Vol. 78
Managing Editor: T. Scheper

Abbreviations

CFR	Code of Federal Regulations
CERCLA	Comprehensive Environmental Response, Compensation, and Liability Act
NPDES	National Pollutant Discharge Elimination System
RCRA	Resource Conservation and Recovery Act
TI	Technical Impracticability
TCLP	Toxic Characteristic Leaching Procedure
US EPA	United States Environmental Protection Agency

1 Introduction

Phytotechnologies are an emerging group of technologies that are expanding in application to a variety of media, site conditions and contaminants. Typical media includes soils, surface water, and groundwater. Depending on the concentrations of constituents and their sources, the media may be considered characteristic hazardous waste or may contain listed hazardous waste.

Phytotechnology projects are being implemented under various laws, rules and regulations throughout the world. Whether phytotechnologies can be utilized at any site or facility is based on the remedial objectives, site conditions, clean up levels that must be achieved within a specified time frame, evaluation criteria, performance criteria, and regulatory approval and reporting. These factors or criteria must be clearly defined and understood by site owners, operators, regulators, stakeholders, and the public in order for phytotechnologies to be successfully implemented. This chapter identifies the regulatory questions or issues that should be addressed when considering phytotechnologies as a remedial technology at a site. Issues presented and discussed are meant to stimulate consideration of the issues germane to implementing phytotechnology projects. This chapter is not meant to establish or interpret federal or state laws or policies.

2 Regulatory Structure for Implementing Phytotechnologies

Phytotechnology projects are likely to be implemented under a variety of environmental laws. In the United States, these projects may be implemented pursuant to federal statutes such as the Resource Conservation and Recovery Act (RCRA), the Comprehensive Environmental Response, Compensation, and Liability Act (CERCLA), Brownfields initiatives, and the Clean Water Act through a National Pollutant Discharge Elimination System (NPDES) permit for point source discharges or to address non-point source pollution. Furthermore, state and local governments have developed voluntary clean up laws which may be applicable as well. Each of these may govern the phytotechnology application. Similar or corresponding laws will be applicable in other parts of the world. The remedial objectives that phytotechnologies are capable of meeting at a site as part of the actions taken under the laws cited above are containment, capping, stabilization, treatment, or a combination thereof. The ability of plant based systems to meet these objectives will depend on site-specific requirements for protection of human health and the environment and the ability of plants to achieve specific performance standards. While site-specific requirements for each regulatory program may vary, basic elements will be common to all. A basic understanding of these regulatory objectives is required to properly formulate a waste management or remediation proposal for presentation to regulators.

2.1 Containment

For sites where wastes or contaminants must be isolated either physically or hydraulically, the regulatory focus will be on the design, performance monitoring, and/or post-closure care monitoring. The goals that will need to be achieved are waste and contaminant isolation to prevent releases into the environment and/or prevention of water infiltration into wastes to minimize or eliminate the formation of leachate. For instance, the design of a vegetative cover must intercept rain and prevent infiltration or it must provide hydraulic control by taking up and removing water after it has entered the subsurface. Performance monitoring must demonstrate that the vegetative cover is effective in meeting design objectives at a designated point of compliance.

Another example of vegetative hydraulic control is the use of deep rooting trees which prevent or minimize dissolved contaminant migration by causing a local groundwater depression through uptake and transpiration. For this type of system, the regulatory focus will be on design and maintenance of the hydraulic barrier, performance monitoring at a point of compliance or other location within the plume for gradient reversal, and the concentration of dissolved contaminants in downgradient monitoring wells. Additional focus will be on whether supplemental hydraulic controls will be needed during periods of plant dormancy, initial tree planting, or periods of high groundwater flow.

2.2 Capping

Plant based systems can be used as alternatives to landfill covers either as evapotranspiration covers (also known as water-balance covers) or phytoremediation covers. Within RCRA, there are separate regulations addressing landfill covers for municipal waste under Subtitle D and for hazardous waste under Subtitle C. The regulations differ in requirements, performance criteria, and monitoring but the intent of both sets of regulations is to isolate waste within an engineered cell that over time, will eliminate or minimize the potential for releases to the environment. Furthermore, these regulatory frameworks also focus on monitoring the performance of the cap during a post-closure care period, typically thirty years. Subtitle D regulations, 40 Code of Federal Regulations (CFR) Part 258, allow approval of alternative landfill cover designs. However, individual state or local solid waste agencies may differ on allowable alternative cover designs, evaluation criteria, and performance requirements. The local solid waste agency should be contacted for area-specific requirements prior to proposing an alternative landfill cover. Approval of alternative landfill covers under Subtitle C is more problematic since 40 CFR 264 Subpart N establishes a permeability engineering standard. There are demonstration projects throughout the United States that are collecting data that may help to establish a performance-based standard, given site-specific weather, geotechnical, and waste conditions. Until these projects generate data demonstrating equivalence, regulators may have to make individual equivalency determinations based on design, performance monitoring, site-specific modeling, and other factors which may arise during the post-closure care period.

2.3 Stabilization

Stabilization measures are often regulatory goals for sites. Stabilization measures may include preventing the release of site contaminants via surface runoff, preventing erosion by wind or water, preventing deep percolation of surface soil contaminants, or protection of workers due to exposure to site contaminants in soil. In some instances, stabilization measures will also act to phytoremediate surface soils. Again, the regulatory focus will be on determining the appropriateness of phytotechnologies at meeting the goals, design, and effectiveness of the stabilization measures at points of compliance.

2.4 Treatment

Treatment, defined by the US EPA pursuant to RCRA rules and regulations in 40 CFR 260.10, includes any activity designed to change the physical, chemical, or biological character or composition of any hazardous waste. The regulatory focus for treatment is how and where treatment will occur and whether wastes will meet treatment objectives, clean up levels, or permit effluent limits. Phytotechnologies that treat wastes, specifically known as phytoremediation, are possible since plants and the rhizosphere organisms interact with waste or effluent to change its physical or chemical nature. The type of media to be treated, in-situ versus ex-situ treatment, and the complexity of the treatment train are a few of the factors that will determine the required degree of regulatory permitting or regulatory involvement.

If the site contaminant to be treated can be translocated into plants, is made more available to organisms because of treatment, or has metabolic degradation byproducts that are more toxic than the original contaminant, then regulators may have questions which will need to be addressed regarding the creation of new exposure pathways and proper management of plant materials that may become toxic or hazardous waste. The concerns of regulators will need to be addressed as part of the phytoremediation design and the short- and long-term monitoring program implemented throughout the life of the project. Understanding treatment mechanisms that are being employed and the fate or transformation of site contaminants within plants will determine: 1) additional measures that may need to be implemented to define performance monitoring for short and long-term maintenance, 2) the scope of any risk assessments, and/or 3) risk mitigation necessary to protect human health and the environment.

2.5 Regulatory Goals Using Phytotechnologies

The ultimate regulatory goal for site remediation is the continued protection of human health and the environment while achieving clean up goals or permit limits. Proponents and regulators must recognize the limitations of phytotechnologies. Otherwise, these goals will not be met. Phytoremediation has been used to remediate a variety of site contaminants. In most cases, plant roots must be in direct contact or within the oxygen diffusion range of the contaminated media. Another limitation of phytotechnologies is the seasonal nature of plant growth which affects transpiration rates, dormancy periods, growth rates, and the amount of time needed to achieve clean up levels. Limits imposed by climatic factors must also be recognized since the length and conditions of the growing season are directly related to growth rates. Also limiting are: 1) contaminant concentrations within media which cannot be phytotoxic, 2) available information about which contaminants are acceptable for phytoremediation and which plants species (either domesticated or native) are capable of degrading, mineralizing, or extracting contaminants, and 3) a lack of performance data demonstrating long-term trends in reduction of contaminant concentration. Rooting depth will also need to be considered as part of the design since the plants which are chosen will dictate the effective rooting depth. Additionally, general site characteristics may also pose limitations since phytotechnologies typically require large surface areas and extensive use of agronomic practices and farming equipment [1].

Given the above limitations of phytotechnologies, proponents and regulators must also consider site-specific factors which may significantly affect the design of the project. Such site-specific factors may include some or all of the following: 1) adequate site characterization to define the distribution of site contaminants throughout the horizontal and vertical soil and groundwater profiles, 2) sources of irrigation water or whether irrigation is necessary, 3) soil and groundwater concentrations of essential nutrients or naturally occurring salts that may affect plant growth, and 4) site preparation and planting methods.

In order to demonstrate compliance with the protectiveness and clean up standards, regulators will require designs that incorporate environmental measurements at points of compliance. Points of compliance vary with environmental media and waste management units. They may be located throughout areas of contaminated soil and groundwater plumes or defined by physical parameters such as vertical planes at downgradient edges of waste management units. Other points of compliance may also be defined depending on site-specific regulations or policies.

Where performance and compliance criteria are specified, consideration must be given to the development of adequate environmental sampling and analysis plans. Sampling and analytical techniques for phytotechnology systems must address the variability of contaminant distribution in three dimensions, potential effects of mixed wastes, composition of contaminants and potential by-products, and the treatment effectiveness. Due to the typically long periods of time required to achieve performance and clean up goals, successful phytotechnology designs will incorporate short and long-term performance monitoring of operations and maintenance as well as analytical approaches for demonstrating compliance with protection standards, contaminant removal, and achievement of site clean up goals. Furthermore, the sampling matrix should be designed to adequately represent the site so that conclusions drawn from the data are reasonably confident. Usually, a statistician with expertise in statistical experimental design and spatial techniques such as kriging and contour plotting should be consulted to help design a tight three-dimensional sampling grid that will enable inferential statistical analyses.

3 Selection and Evaluation Standards for Phytotechnologies

Phytotechnologies represent another alternative to technologies already available to clean up contaminated sites. Therefore, they should be judged as appropriate based on the same criteria. However, there may be certain sites that exist where no alternative is technically practicable. Under these circumstances, the use of phytotechnologies may be reasonable due to certain ancillary benefits as well as the long-term nature of the remedy.

3.1 Selection Criteria and Evaluation Factors

When phytotechnology solutions are presented to regulators for consideration in cleaning up an impacted site, they will be evaluated using specific standards and factors. These will be identical to the prescribed expectations designated for any proposed remedy. At CERCLA and RCRA hazardous waste sites, the standards and factors for remedies are very similar. Technologies prescribed for contaminated sites shall, at a minimum, meet the following four standards:

1. Be protective of human health and the environment – Phytotechnologies will have to be shown to eliminate, reduce, or control threats to public health and the environment.
2. Attain media clean up standards – Phytotechnologies will need to have documentation demonstrating the ability to meet media clean up goals within specified time frames.
3. Control the sources of releases – Phytotechnologies will have to demonstrate their ability to reduce or eliminate, to the extent practicable, further releases of hazardous wastes.
4. Comply with applicable and relevant standards for the management of wastes – Phytotechnologies will have to be designed to comply with applicable federal, state, and local waste management requirements.

After meeting these four standards, remedies will be evaluated using the following factors, as appropriate:

1. Long-term reliability and effectiveness – Phytotechnologies will be evaluated for their long-term reliability and effectiveness along with the degree of certainty that the remedy will prove successful.
2. Reduction of toxicity, mobility, or volume – Phytotechnologies will be evaluated for their ability to reduce the toxicity of contaminants, the ability of contaminants to move in the environment, or the amount of contamination present, based on remediation goals or regulatory standards.
3. Short-term effectiveness – Phytotechnologies will be evaluated by the length of time needed to implement the technology as well as whether the phytotechnologies are able to reduce risks over the short-term including during the implementation phase.
4. Implementability – Phytotechnologies will be evaluated by the difficulty or ease of implementation.
5. Cost – Phytotechnologies will be evaluated using the net present value of capital and operation and maintenance costs.
6. State and community acceptance – Phytotechnologies may be part of an enforceable document such as a legal order, record of decision, or permit. These may require public comment periods and hearings, a letter acknowledging the acceptance of public comment, a responsiveness summary or other documentation that provides a response to comments, and a listing of the reasons for an agency decision.

When site clean up remedies are presented to regulators and evaluated using these standards and evaluation factors, rarely does a specific remedy stand out above all other remedies. However, all must generally meet the first four criteria to even be considered. In general, phytotechnologies are capable of meeting these remedy selection standards. Like all remedies, phytotechnologies tend to have inherent strengths and weaknesses that depend on site conditions and contaminants. For complex sites with varied wastes and geological conditions, it is likely that the best remedy will involve multiple technologies where the strength of each technology can be optimized.

3.2 Technical Impracticability

The US EPA has Technical Impracticability (TI) waivers available for sites. These may be granted where unique waste and site characteristics yield no remedy that is technically practicable for clean up of soils or groundwater at the time of remedy selection. US EPA criteria for determining that a site is a candidate for a TI waiver are found in the Guidance for Evaluating the Technical Impracticability of Groundwater Restoration Interim Final Directive 9234.2–25 (see reference [2]). To receive a TI waiver, sites will have to demonstrate, among other things, containment of plumes and control of access to the site. Once the TI waiver is received, a site will not be expected to meet final clean up goals if it is a CERCLA site and will be deferred from meeting timeliness goals if it is a RCRA site. This combination of containment, access control, and relief from timeliness requirements may make TI sites good candidates for use of phytotechnologies where long time frames are needed for achieving clean up goals. However, it should also be noted that TI sites also provide challenges in applying phytotechnologies due to the inherent limitations of the technology and complexity of site conditions. Proponents and regulators may wish to use existing documents in deciding whether phytoremediation is appropriate at any given TI site such as the Introduction to Phytoremediation (see reference [3]) or Phytoremediation Decision Tree (see reference [4]).

At TI sites, it may be possible for phytotechnologies to enhance or complement the final overall remedy. Phytotechnologies are not just for beautification, but act to stabilize soils, reduce surface runoff, and reduce dissolved contaminants in groundwater. Sites with TI waivers, by their nature, will require long-term maintenance and monitoring. These sites may require periodic (five year) reviews of the remedy, provide an opportunity to conduct long-term studies regarding the list of applicable contaminants, and to evaluate the long-term performance of phytotechnologies. These sites also afford opportunities for expansion to full-scale application outside of any regulatory requirements. Such sites may generate information on long-term performance as well as maintenance and monitoring costs.

4 Regulatory Obstacles to Implementing Phytotechnologies

The regulatory issues that pose challenges to developing phytotechnology projects include adequate site characterization, points of compliance, waste determination requirements, monitoring methods, demonstrating achievement of remediation goals, and risks to ecological receptors. These regulatory issues only become obstacles if proponents and regulators do not incorporate these concepts as part of the design of the phytotechnology project, the long-term performance monitoring, and the sampling and analysis plan for determining whether remediation goals have been achieved.

4.1 Site Characterization

Regulators are concerned with assuring appropriate characterization of contaminants across a site. This characterization is essential for determining the effectiveness of plant based remediation designs and their ability to meet remediation goals. Simple sites with limited releases of contaminants and low geological complexity will require less characterization than more complex sites. A full and comprehensive characterization of the variabilities in surface and sub-surface contamination distributions and concentrations is essential for determining: 1) base-line conditions, 2) phytotoxic conditions, 3) the potential for plant root systems to reach and remove contaminants from soils and groundwater, and 4) demonstrating the achievement of clean up goals. The degree of characterization and type of information collected should also be tailored to satisfying future points of compliance or performance requirements specific to the applied technology. For instance, where plant hydraulic barriers with or without remediation capabilities are contemplated, site characterization should include knowledge of where the leading edge of the plume is located in order for proper placement of the barrier in the path of flow. Furthermore, the range of contaminant concentrations the vegetation is likely to intercept should be known in order to determine potential toxicities. The performance criteria for a plant-induced hydraulic barrier may include a comparison of concentrations both up- and downgradient over time or a comparison to background concentrations over time.

After phytotechnologies are implemented, site conditions will change making meaningful comparisons between pre- and post-remedy implementation difficult to impossible. Progress toward site remediation goals for most phytotechnologies is determined by periodic comparisons of current conditions to baseline conditions. Therefore, the lack of comprehensive initial site characterization data will confound future analyses designed to demonstrate progress and the achievement of remediation goals.

4.2 Demonstrating Progress at Points of Compliance

Often overlooked in the preparation of phytotechnology projects is the importance of properly characterizing the distribution of contaminants in three dimensions across a contaminated site. Achievement of remediation clean up goals will be difficult to measure if baseline site characterization data is lacking. Inadequate site characterization may result in an incomplete understanding of contaminant concentrations, distribution, and variability. Typically, the points of compliance designated in the site remediation goals for soils and groundwater are located throughout the impacted areas. Since points of compliance are three-dimensional in nature, the distribution of contaminants must also be understood in three dimensions to successfully demonstrate the achievement of site remediation goals.

In some governmental jurisdictions, current and future land uses will determine clean up levels while in others, clean up levels are set to unrestricted or residential use levels. These unrestricted use levels consider background concentrations of metals in soils and groundwater and will additionally use the Safe Drinking Water Act Maximum Contaminant Levels (where available) as clean up goals. Site-specific conditions as well as current or future land use will determine the depth to which the point of compliance will be applied. For groundwater, like soils, site-specific conditions will determine where the compliance point will be located. Typically located throughout the plume of contamination, groundwater compliance can also be measured at the edge of a plume, property boundary, or control area that may be restricted by enforceable institutional control mechanisms. Consultation with state and federal regulatory agencies overseeing the site remediation is recommended to determine how points of compliance will be defined.

4.3 Transport of Contaminants into Plants

Phytotechnologies that are implemented at sites where the treatment of soils or groundwater is occurring will potentially result in site contaminants moving into plant tissues. Regulators will be concerned that this translocation of contaminants, or in some instances, transformation of site contaminants to more toxic metabolic by-products, may pose waste management concerns. For sites where plants are being used to treat soils or water containing characteristic hazardous wastes as defined in 40 CFR Part 261, Subpart C, movement of these constituents into plants may cause the vegetation to also become a characteristic hazardous waste. This depends on the constituent concentrations in the plant which are considered hazardous if they meet or exceed toxicity characteristics listed in 40 CFR Part 261.24. The toxicity characteristics are based on measuring constituent concentrations in plant tissues using extracts derived by the Toxic Characteristic Leaching Procedure (TCLP). Not all plants grown on characteristic hazardous wastes will become a characteristic hazardous waste. As of yet, laboratory and field data have not demonstrated significant accumulations of constituents at concentrations which would be considered characteristic hazardous waste levels. However, future innovations in the field or as of yet undiscovered species with remedial capabilities may yield plants with such capabilities. Therefore, before disposal of potential plant wastes, it is recommended that representative sampling and analysis of plant tissues be performed to determine whether the concentrations achieve hazardous waste levels.

Phytotechnologies may also be implemented on sites where plants will be treating listed hazardous wastes according to 40 CFR Part 261, Subpart D. Based on US EPA regulations found in 40 CFR Part 261.3(c)(2)(i), also known as the "derived-from" rule, residues resulting from the treatment of listed hazardous wastes are also considered to be listed hazardous wastes themselves. It is, therefore, possible to conclude that plants treating listed hazardous wastes may become listed hazardous wastes themselves if the constituents are found in plant tissues. These waste residues can be delisted after a determination is made and presented that a treatment residue is no longer a hazardous waste. The US EPA has issued guidance on the information that must be presented as part of a delisting petition in 40 CFR Part 260.22 (b). If such an interpretation is made then proponents and regulators must make decisions early in the design of the phytoremediation project that describes how plant wastes will be managed and potentially delisted.

Official guidance has yet to be issued by the US EPA on the applicability of the derived-from rule to plants used for remediating listed hazardous waste sites. However, state or local hazardous waste programs may independently write guidance and policies on the applicability of the derived-from rule to phytoremediation plant wastes. These could exclude plants from being considered listed hazardous wastes. Proponents expecting to use phytoremediation on sites contaminated with listed hazardous wastes are encouraged to consult with the local regulatory agency for their determination on proper management of potential plant wastes.

If plant wastes are not considered either a characteristic or listed hazardous waste, they may be managed as solid waste. Solid waste management options may include composting, incineration, or air drying the plants prior to placement in a municipal solid waste landfill. Disposal of plant wastes will be a regulatory focus that should be considered early on. This should be part of the phytoremediation design and be included as part of the long-term monitoring or maintenance plan for the site.

4.4 Ecological Impacts

In addition to the various regulatory concerns, ecological issues could also pose obstacles to implementing phytotechnologies. Waste constituents that are transported into plant tissues may make contaminants more available to other organisms. This will potentially create new exposure pathways to ecological receptors. Specifically, animals may be attracted to the area for food, nesting, or other activities. Another route of exposure is the potential transfer of contaminants to the air through phytovolatilization. This occurrence may require air monitoring. The recommended response to both of these issues is to prepare an ecological risk assessment as part of the phytotechnology design. Guidance for developing ecological risk assessments have been developed by the US EPA and can be found in the Guidelines for Ecological Risk Assessment (see reference [5]) or the Ecological Risk Assessment Guidance For Superfund (see reference [6]). After evaluating the ecological risks, appropriate control measures can be developed to mitigate those risks considered unacceptable. Alternatively, the remediation design can be altered to minimize or eliminate these risks.

5 Long-Term Site Maintenance and Monitoring

Phytotechnologies, like any other clean up system, will require monitoring and maintenance. The survivability and growth rate of plants directly affects the effectiveness of the system and should be monitored closely during the life of the project. Monitoring must be done to assess performance as well as to optimize or maintain the efficacy of the phytotechnology projects.

5.1 Maintenance Activities

Examples of maintenance activities that optimize phytotechnologies include adjusting fertilization and irrigation rates, adding other soil amendments, tracking climatic conditions, mowing or pruning plants for optimal growth, controlling weeds and pests, managing plant diseases, and replacing or replanting the vegetation [1,3]. These types of long-term maintenance of phytotechnology projects are essential parts of the design. Without this attention to maintenance details, meeting performance objectives or remediation goals may be unattainable. Regulators will likely view maintenance as part of the design of the project and will require operation and maintenance to be addressed in a separate manual or checklist for the site. Maintenance activities that need to be reported to regulators will vary with the frequency and type of activity and will need to be conducted over the life of the phytotechnology project.

5.2 Monitoring Phytotechnology Systems

In addition to maintenance, plant surveys will be need to be conducted to assess plant health, growth and survivorship, vegetative cover density, rooting depth, root-to-shoot ratios, leaf density measurements, new shoot growth, and other plant-related parameters. These may become part of the performance monitoring. Plant tissue samples such as leaves, stems, or shoots may also need to be collected to determine if site contaminants are accumulating within the tissues. This information may be needed as part of an environmental risk assessment. The methods and frequency of monitoring should be identified as part of the design while results should be reported periodically to regulators as part of performance monitoring.

Soil and groundwater sampling will also likely be required as part of site performance monitoring or maintenance. Remediation goals and performance monitoring may include demonstrating continuous reduction of site contaminants in the impacted media. Ultimately, these may be needed to demonstrate the achievement of site clean up goals over the entire three-dimensional treatment zone. The design must include appropriate sampling and analysis to characterize site contaminant concentrations both spatially and as a function of time. Other monitoring requirements may need to determine whether site contaminants are likely to migrate to other media or other areas. Again, regulators will expect this information to be part of the design and the long-term monitoring plan. Furthermore, the results will need to be periodically communicated in performance monitoring reports.

6
Achieving Performance or Remediation Goals with Phytotechnologies

Phytotechnologies are generally long-term remedies. Therefore, models and sound statistical analyses can be used to demonstrate the actual achievement or, in the interim, as a measure of progress towards the eventual completion of the remedial goals. These goals will be dependent on the future land uses envisioned for the site and will be dictated by the applicable regulatory framework. Furthermore, these analyses can be used to trigger potential contingency plans should the phytoremedy prove unable to meet the desired end points.

6.1
Time to Achieve Remediation Goals

The length of time to reach clean up levels or remediation goals for many site contaminants is unknown and difficult to predict with any certainty. Thus, if warranted by site conditions, proponents and regulators must allow reasonable time to evaluate phytoremediation that may extend into years or decades. Sites where phytoremediation has been implemented as part of a TI waiver may provide the greatest opportunity to evaluate the long-term effects of phytotechnologies leading to better estimates about contaminant reduction rates, effects on the surrounding environment, or risk to ecological receptors.

Groundwater transport models are frequently used in calculations which predict the amount of time needed to remediate a site. Where models are used, regulators will require validation with site-specific data. These will also be used to calibrate assumptions about discreet parts of the overall phytotechnology system. Furthermore, the data needed to demonstrate performance should be identified early in the design process. This requirement is needed in order to calibrate the model as well as insure that the appropriate data is collected. Conclusions based on models that have not been validated or where essential data has not been gathered will be compromised.

6.2
Statistical Characterization of Contaminant Removal by Phytoremedies

Statistical analysis is required to make determinations regarding whether progress is being made toward site remediation goals. Due to the extreme variability of contaminant concentrations at most sites and the fact that phytotechnologies will not degrade or remove all contaminants at the same rate, demonstrating progress toward site remediation goals will not be a straightforward comparison at most sites. Progress of phytoremediation at sites with normal distributions of contaminants and simple site conditions will most likely be evaluated with parametric statistical tests. However, sites with non-normal distributions of contaminants and complex site conditions will require the incorporation of a rigorous statistical design for evaluating long-term monitoring or performance data. Proponents of phytoremediation projects are encouraged to consult with a statistician for appropriate design, site characterization, sampling plan, and data analysis [7].

For all sites, statistical analysis, which will be used to demonstrate compliance with performance or clean up goals, needs to be incorporated as part of the performance monitoring that documents the change of contaminant concentrations with time or spatially. This statistical design will be based on site characterization data which should have adequately characterized the three-dimensional distribution of contaminants in soils and/or groundwater. The statistical design should include quantifying the variability of site characterization and performance data while allowing for valid inferences to be made. This includes whether clean up levels or expected contaminant reduction levels are achieved over time [7]. Sampling and statistical methods for demonstrating the achievement of groundwater standards at RCRA sites will include monitoring as prescribed in 40 CFR Part 264 Subpart F.

6.3 Achievement of Remediation Goals

A demonstration that site clean up goals have been achieved must incorporate the inherent variability of soil conditions, plant remediation rates, and the range and distribution of contaminant concentrations at the site being remediated. The data needed for such a demonstration will be generated pursuant to the site sampling plan which must address this variability while incorporating points of exposure and future site uses. For soils, the points and areas of exposure will depend on current and future land use. If land use scenarios include residential use or the party responsible for the site intends to eventually "walk away" from the site afterwards, clean up levels, sampling density, and locations of soil samples will be more conservative than for industrial, commercial, or recreational land uses. Similarly, for groundwater, the points of exposure and clean up levels will also depend on current and future land use. However, clean up levels may also be based on non-degradation or classification standards. Typical groundwater standards are generally fixed and based on levels derived from drinking water standards, background, or potential risks.

Statistical tests can also be used to analyze sampling data to evaluate whether clean up levels have been achieved. Proponents of phytotechnology projects should work with regulators early in the project development to identify the precision needed to answer statistical hypotheses regarding the likelihood of contamination levels exceeding remediation goals. In order to specifying the number and type of samples to be taken, an adequate sampling plan will incorporate the required degree of precision, the statistical type of contaminant distribution (derived from baseline sampling during initial site characterization), and the spatial variability of contaminants. The type of statistical distribution associated with the site contamination will determine the appropriate statistical tests for analyzing site data. Using composited soil samples may be useful in reducing the variability of data obtained from a site; however, care should be taken not to mask high contaminant concentrations or hot spots with samples taken from low or near-background concentrations. The use of composite soil samples or in conjunction with standard grab samples should be cleared with the regulatory project manager. Furthermore, this masking can dampen the spatial variability of contaminants and introduce errors into the statistical design.

6.4
Contingency Plans

Before approving a phytotechnology project, regulators will consider the consequences if there is a potential that the technology is unable to achieve site remediation goals or adequately protect human health and the environment. Thus, regulators may require a contingency plan as part of the design. The contingency plan should identify possible failure mechanisms or examine how performance monitoring will reveal events that require the implementation of the contingency plan. For sites that have a TI determination, a five year review is required based on the TI guidance developed by the US EPA [2]. Proponents and regulators may structure the five year review of the phytotechnology project using performance monitoring that ensures reliability, protectiveness, and the achievement of overall site restoration goals

7
References

1. ITRC (2001) Phytotechnology Technical and Regulatory Guidance Document, Interstate Technology and Regulatory Cooperation Phytotechnology Work Group
2. EPA (1993) Guidance for Evaluating the Technical Impracticability of Ground-Water Restoration Interim Final, EPA Directive 9234.2-25
3. US EPA (2000) Introduction to Phytoremediation, EPA National Risk Management Research Laboratory, EPA Document # 600/R-99/107
4. ITRC (1999) Phytoremediation Decision Tree, Interstate Technology and Regulatory Cooperation Phytoremediation Work Group
5. US EPA (1999) Guidelines for Ecological Risk Assessment, EPA Document # 630/R-95/002F
6. US EPA (1997) Ecological Risk Assessment Guidance For Superfund, EPA Office of Solid Waste and Emergency Response Document # 540-R-97-006, EPA Directive 9285.7-25
7. Schwab AP (2000) Phytoremediation of Petroleum Contaminated Soils: Field Guide, EPA Great Plains-Rocky Mountain Hazardous Substance Research Center, Kansas State University, Manhattan, KS

Received: November 2001

Author Index Volume 51 – 78

Author Index Volumes 1 – 50 see Volume 50

Subject Index

Printing (Computer to Film): Saladruck Berlin
Binding: Stürtz AG, Würzburg